Hauke Hinsch

Entangled Networks of Semiflexible Polymers

Hauke Hinsch

Entangled Networks of Semiflexible Polymers

Tube Properties and Mechanical Response

Südwestdeutscher Verlag für Hochschulschriften

Impressum / Imprint
Bibliografische Information der Deutschen Nationalbibliothek: Die Deutsche Nationalbibliothek verzeichnet diese Publikation in der Deutschen Nationalbibliografie; detaillierte bibliografische Daten sind im Internet über http://dnb.d-nb.de abrufbar.

Bibliographic information published by the Deutsche Nationalbibliothek: The Deutsche Nationalbibliothek lists this publication in the Deutsche Nationalbibliografie; detailed bibliographic data are available in the Internet at http://dnb.d-nb.de.

Verlag / Publisher:
Südwestdeutscher Verlag für Hochschulschriften
ist ein Imprint der / is a trademark of
OmniScriptum GmbH & Co. KG
Heinrich-Böcking-Str. 6-8, 66121 Saarbrücken, Deutschland / Germany
Email: info@svh-verlag.de

Herstellung: siehe letzte Seite /
Printed at: see last page
ISBN: 978-3-8381-1149-0

Zugl. / Approved by: München, Ludwig-Maximilians-Universität, Diss., 2009

To Whom It May Concern

Contents

List of Figures

Chapter 1

Introduction

1.1 Biological Physics and Polymer Science

Biopolymers and polymeric materials play a crucial role in many aspects of microbiology and biological physics. Probably the most prominent example of a biopolymer is DNA that has become the figurehead of this field of research. Besides the storage of genetic information, biopolymers are involved in such a variety of vital processes that a functional cell cannot be thought of without them. They form versatile materials that define mechanical properties of the cell, facilitate transport through the crowded intra-cellular environment, separate the nucleus during mitosis and actively change their structure to allow for cell motility.

Most of the cell's biopolymers are organized in the cytoskeleton that is often described as a scaffold, but yet performs a variety of functions that go well beyond the mere provision of structural support. A microscopic picture of the cytoskeleton of a mouse embryo is shown in Figure 1.1. The main components of the cytoskeleton are commonly distinguished by their size: microfilaments like filamentous actin (F-actin), intermediate filaments like vimentin or keratin, and microtubules. F-actin is formed by polymerization of globular actin into a double stranded filamentous structure of around 6 nanometer diameter [2] and is most abundant near the cell membrane (see Figure 1.1). Here it forms an active network with the help of actin binding proteins [3] as for example the branching protein Arp2/3 [4] (for a microscopic picture see Figure 1.2) and allows for cell motility and the transduction of mechanical signals. Cellular motion is realized by the extension of a leading edge - the lamellipodium - through a dynamic growth of the actin network [5]. Remarkably this complex motility seems to be an autonomous feature of the actin network as it is also observed in cells that lack nuclei and microtubules [6]. In muscle cells F-actin serves as a structure that permits myosin motor proteins to generate forces [7, 8]. Microtubules consist of several protofilaments that are arranged into a hollow structure (depicted in Figure 1.3) with a diameter of some 20 nm [9] . They do not only provide structural support, but also act as tracks along which motor proteins like dynein and kinesin transport cargo [10, 11, 12, 13]. Furthermore, microtubules play a key role in mitosis forming the

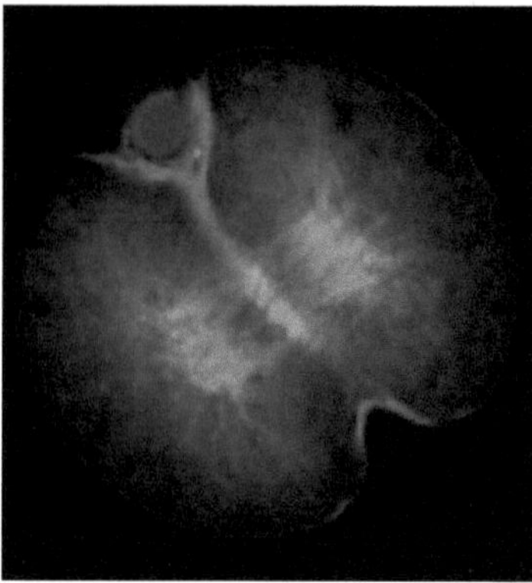

Figure 1.1: Cytoskeleton of a mouse embryo during cell division reproduced from [1]. Fluorescent picture of microtubules (green), actin (red) and DNA (blue). Actin is most abundant near the cell membrane and DNA is located in the nuclei between which the mitotic spindle formed by microtubuli can be seen.

spindle that segregates chromosomes during cell division [14, 15]. They are also important components of cilia, flagella and other bundle-like structures [16].

The cytoskeleton is thus a complex network that has to perform a variety of seemingly contradictory tasks inside the cell. On the one hand it has to provide structural support and mechanical stability - features that call for a rigid material. But on the other hand also a flexibility is needed that the cell requires for needs like cell division or cell motility. The fact that the cytoskeleton successfully complies with these requirements can be traced back to different levels: the features of its basic polymeric constituents, the interplay of these biopolymers in networks, and the dynamic processes induced by polymerization and binding proteins. All these levels exhibit challenging physical questions and are essential for the understanding of the properties of eucaryotic cells.

Also apart from cellular biology, biopolymers are ubiquitous. Fibrin ensures blood coagulation, collagen is a main constituent of biological tissue and cellulose is one of the most abundant organic compounds on earth. Since many natural materials are thus polymer networks, it is not surprising that also the technical sciences have tried to develop useful materials from polymers. This tradition goes back to the fabrication of paper from cellulose and with the advent of synthetic polymers has created a vast variety of materials like e.g. plastic or rubber.

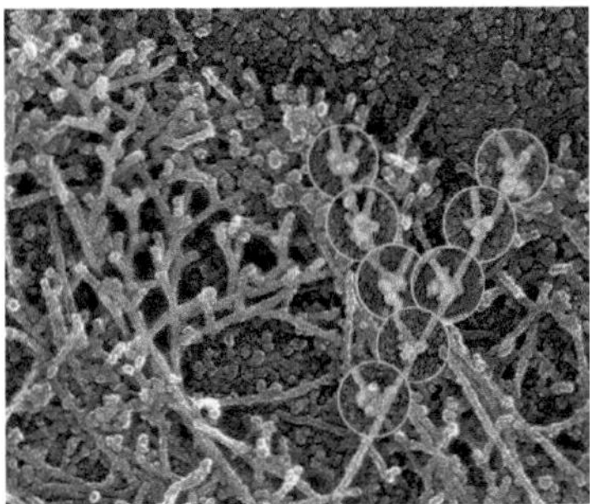

Figure 1.2: Electron Microscopy picture of a F-actin network in a keratocyte lamellipodium reproduced from [4]. The characteristic 70 degree angle at the branching points (yellow circles) is caused by the binding protein Arp2/3.

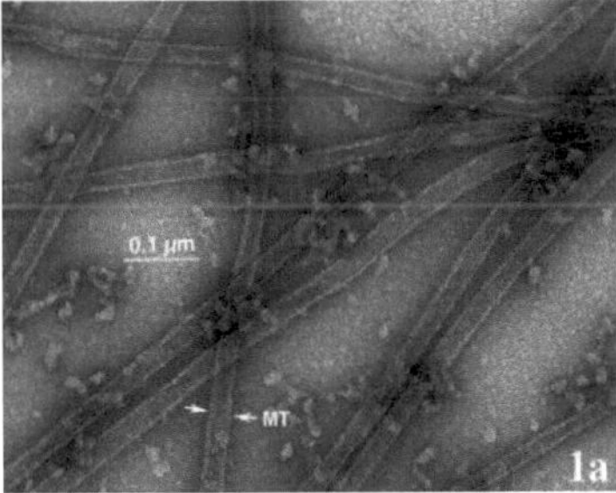

Figure 1.3: Transmission electron microscopy picture of microtubules from rat tubulin reveals a hollow structure that gives rise to a high stiffness. Reproduced from [17].

1.2 Single Polymers

To gain insight into the physics of polymer networks it is essential to first look at their building block - the single polymer. Take for instance the elasticity of rubber: its physical origin and its dependence on external parameters can be described by looking at a single flexible polymer from which the material is build by crosslinking. In a solvent at finite temperature the polymer is subjected to permanent collisions with solvent molecules that effortlessly change the polymer's shape. In thermodynamic equilibrium a flexible polymer can thus be treated as a random walk with a vanishing step size (see Figure 1.4). Consequently, the mean value of the end-to-end distance vanishes. If however a force is applied to the ends and they are drawn apart, the complete polymer stretches out and the number of accessible configurations of the chain is reduced until it finally reaches one in the completely stretched state. The restoring force of the polymer is thus of entropic nature and increases with increasing temperature. The same result can also be observed for a macroscopic sample of rubber.

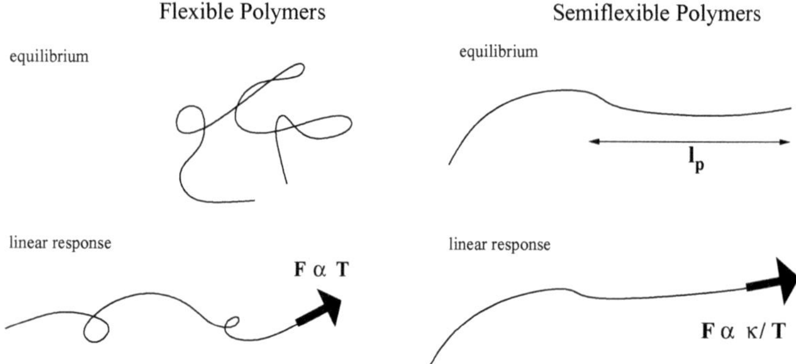

Figure 1.4: (left) Flexible polymer in an equilibrium and in a stretched state. The response is of entropic nature with a spring constant proportional to temperature. (right) The local bending of a semiflexible polymer costs an amount of energy proportional to the square of the local curvature. This results in an additional length scale - the persistence length - that can be seen as the length on which the polymer appears straight. The linear response is also entropic in origin but exhibits a different scaling law.

While the behavior of flexible polymers is dominated by entropic effects, semi-flexible polymers like most biopolymers additionally have internal energy contributions. These arise upon bending of the polymer and introduce an additional length scale - the persistence length l_p. The persistence length emerges as a result of the competition of internal bending energy and entropy. This is reflected in the relation to the local bending stiffness κ that takes on $l_p = \kappa/k_\mathrm{B}T$ where k_B is the Boltzmann constant and T the temperature. It signifies that now the thermal energy transfer from the solvent is resisted by the polymer's stiffness. Phenomenologically, this length can be described as the average distance on which the polymer appears straight. Depending on the molecular structure of a biopolymer but also on the cellular environment this length strongly differs. Microtubules for instance are rather stiff with persistence length of several millimeters [18, 19] and therefore appear completely rigid on the length scale of the cell (see Figure 1.3). F-actin however exhibits a lower bending stiffness resulting in a persistence length of approximately 10 μm for unstabilized filaments [20] and up to 20 μm for phalloidin stabilized filaments [21, 22, 23]. The persistence length is thus of the same magnitude as typical polymer lengths L in the cellular environment or in typical in vitro preparations making F-actin a textbook representative of a semiflexible polymer.

From the theoretical perspective a semiflexible polymer is conventionally described in terms of the worm-like chain model [24, 25] where the polymer's configuration in space is modeled by an inextensible line $\mathrm{r}(s)$ parameterized by the arc length s. Derivation by arc length yields the tangent $\mathrm{t}(s) = \partial \mathrm{r}(s)/\partial s$ and repeated derivation can be used to define a local curvature $|\partial \mathrm{t}(s)/\partial s|$. The energy cost required to bend the semiflexible polymer is then proportional to the square of this curvature. The resulting Hamiltonian of a worm-like chain of length L is hence given as

$$H_{\mathrm{WLC}} = \frac{\kappa}{2} \int_0^L ds \left(\frac{\partial^2 \mathrm{r}(s)}{\partial s^2} \right)^2 . \tag{1.1}$$

While this Hamiltonian looks rather tractable at a first glance, its analysis is considerably complicated by the fact that the inextensibility of the polymer has to be taken into account by the condition $|\mathrm{t}(s)| = 1$ locally. Therefore, only few analytical results exist for the worm-like chain. One of these is the tangent-tangent correlation function $\langle \mathrm{t}(s)\mathrm{t}(s') \rangle = \exp(-|s - s'|/l_p)$ that illustrates the significance of the persistence length as the length scale on which the polymer's directional information is conserved, i.e. where it remains rather straight. Besides from moments of the end-to-end distribution [25, 26] further analytic results are only available in limiting cases one of which is the weakly bending rod approximation [27]. Here the polymer is treated as a straight rod with small transverse bending fluctuations. This approach allows for instance to calculate the radial distribution function of the end-to-end distance [28] or the linear response coefficients [29]. While this response is again entropic in its nature as undulations are pulled out by a parallel force, its dependence on temperature is completely different from flexible polymers and scales inversely with temperature.

In order to investigate properties beyond the weakly-bending regime or to corroborate simplifying theoretical models, a convenient tool are computer simulations of polymers. While for equilibrium properties Monte-Carlo simulations have proven useful, also non-equilibrium properties and dynamics can be studied by Molecular Dynamics [30, 31].

1.3 Networks of Semiflexible Polymers

After having introduced semiflexible polymers and their properties we can proceed and have a look at their assemblies. Since already the theoretical treatment of a single polymer can be a challenging task for the theoretical physicist, it is not surprising that a complete network with its mere number of constituents and a variety of possible interactions is an even more complex problem. The possible interactions in the cytoskeleton of the eucaryotic cell range from simple hard-core interactions over potentials and chemical bonds up to biological motors and biochemical processes like polymerization and depolymerization. In an attempt to simplify the system from the theoretical point of view and with the goal to obtain reproducible experimental systems in vitro, it is often useful to categorize the networks by their interactions: we distinguish entangled networks, crosslinked networks and active gels. Entangled or physical networks are pure solutions of polymers without the presence of any binding proteins or ATP. In F-actin solutions a stabilizing agent like phalloidin is usually added to prevent depolymerization and to obtain a stable solution of filaments of constant length. The only interaction between the polymers in such a preparation is of topological nature and the constituents can effortlessly slide past each other but are not allowed to cross. The resulting physics of these networks will be discussed in more detail in the next chapters. Crosslinked or chemically networks however, are composed of polymers and one or more binding proteins that form permanent links between the single polymers as can be seen in Figure 1.5. The resulting mechanical properties bear resemblance to classic rubber [32] but depend sensitively on the detailed nature of the crosslinker. For instance it has been found that actin filaments may completely organize into bundles [33, 34, 35] resulting in characteristic differences to the "Mikado" models of randomly crosslinked filaments [36, 37]. Finally, the term active gels is used to subsume those polymer assemblies that undergo dynamic changes under the consumption of ATP e.g. by molecular motors. While these networks come closest to meet the cell's constant demand for change and adaption to its environment, their investigation is obviously the most challenging. However, recent work has provided insight into the adaption of mechanical properties by molecular motors [38, 39, 40] and into cell motility based on dynamic polymerization [41, 42].

The experimental possibilities to investigate biopolymers are manifold: besides standard imaging techniques, dynamic light scattering can be used to obtain properties like the persistence length or other correlation functions directly from thermally fluctuating solutions [43, 44, 45]. Also direct visualization of single polymers in solution is possible by fluorescent labelling [46, 47, 48] and has been used to observe the localization of single polymers to tube-like regions by surrounding polymers as depicted in Figure 1.6. The mechanical properties of networks are characterized by rheological methods. In the classical macroscopic or bulk rheology a frequency dependent oscillating strain is applied to a sample and the resulting stresses are measured to obtain the modulus [49, 50]. Recently, microrheological methods have attracted more interest: these include actively oscillated magnetic beads [51] or video-microscopic measurements of passive probes [52, 53].

From the theorist's point of view the mere number of constituents and degrees of freedom in a polymer network obviously calls for the toolbox of statistical mechanics.

The essential simplification to render the system tractable will usually be the reduction to a self-consistent single polymer model. The most successful description of entangled networks of semiflexible polymers in this realm is the famous "tube model" pioneered by de Gennes [54] and Doi and Edwards [55] and will be introduced in the next chapter.

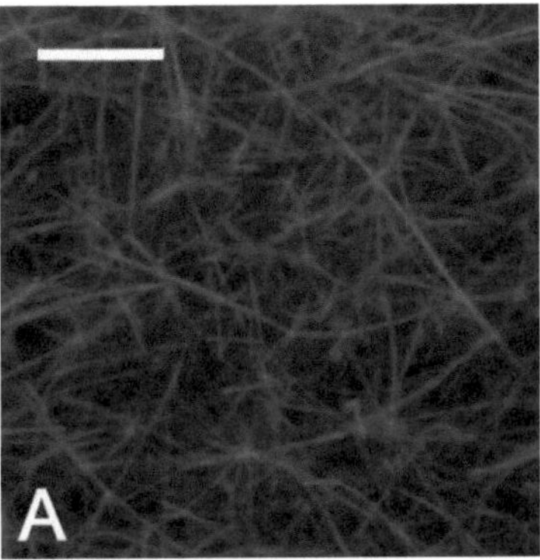

Figure 1.5: Confocal image of a network of F-actin crosslinked by fascin from [56]. The scale bar denotes 10 μm.

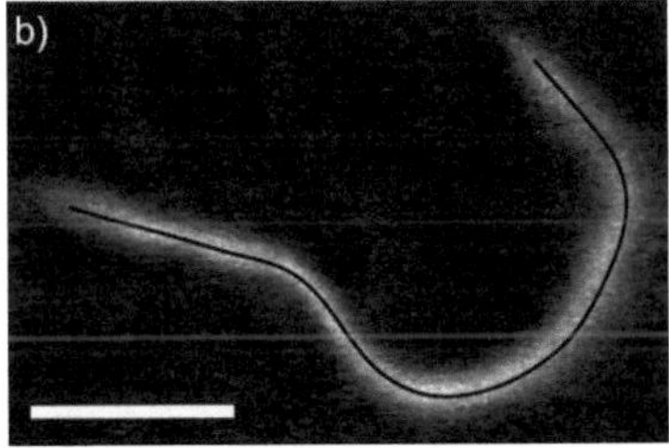

Figure 1.6: Localization of a fluorescently marked test polymer in an entangled F-actin network to a tube-like region. The image is obtained by overlaying single fluorescence microscopy pictures recorded during several minutes [57]. The scale bar denotes 5 μm and the black line represents the tube center.

Chapter 2

Entangled Networks

The essential step needed to successfully grasp the complexity of a polymer network with its enormous number of degrees of freedom is the simplification to a single polymer description. The strategy is to focus on a single polymer, describe the interaction with its neighborhood and extrapolate from the properties of this test polymer to the features of the complete macroscopic material. However, even restricting the analysis to the immediate environment of our probe filament can amount to a challenging task as can be perceived from Figure 2.1. Here the neighboring polymers act as hard-core constraints that limit the available phase space of the test polymer in a complicated topology that on top of it dynamically changes as every constituent of the network undergoes thermal fluctuations. The seminal idea of de Gennes [54] and Doi and Edwards [55] was to reduce this complicated phase space topology to a static environment with a tractable geometry. They assume that the combined effect of all fluctuating obstacle polymers in the direct environment of a specific probe filament can be condensed into an impenetrable, cylindrical tube that follows the average contour of the encaged polymer. Despite its simplicity, this famous "tube model" has proven to correctly predict a multitude of observables at least at a qualitative level. For instance, it faithfully describes how transverse undulations of a polymer are suppressed by its environment giving rise to a free energy penalty or how diffusion of the polymer only occurs along its own contour resulting in strongly modified transverse, reptational and rotational diffusion coefficients. The tube model essentially amounts to a mean-field approximation of the probe filament's direct neighborhood by a virtual tube potential. This potential can be taken to have hard walls but is also often assumed to feature an harmonic form. The parameter that has to be adapted in order to provide a self-consistent description of the network is the tube's diameter or the harmonic potential's strength. Before we proceed to recapitulate scaling law predictions for equilibrium properties of the tube and non-equilibrium properties of the complete network we shall take another look at the length scales involved in the system.

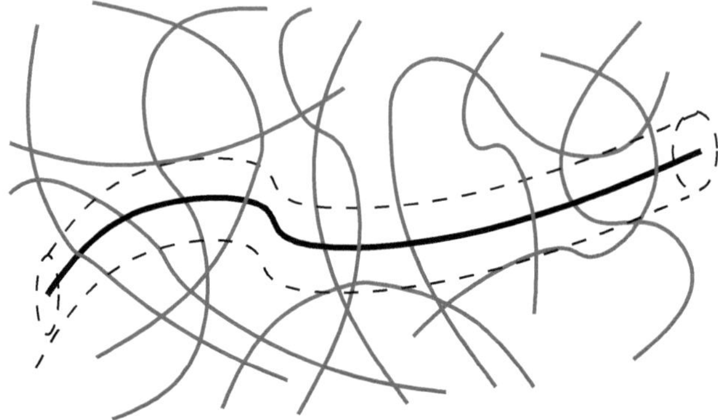

Figure 2.1: The effect of all surrounding polymers (red) that hinder the transverse displacement of a test polymer (black) is described by a hypothetical tube (dashed).

2.1 Equilibrium - Length Scales and Tube Diameter

The first length scale that initially comes to mind is of course the polymer's total contour length L. For classical semiflexible polymers like F-actin in the cellular environment or in conventional in vitro preparations this length is on the order of the persistence length l_p. Since these two length scales describe characteristic properties of the single polymer we subsume them under the notation

$$l_{\text{single}} := L \approx l_p \ . \tag{2.1}$$

Now, we are interested in an additional length scale describing the structure of the network. Here, the most obvious parameter is the number density of polymers ν that is proportional to the monomer concentration c. With increasing density, the space available to an encaged probe filament will evidently diminish as the size of network void spaces reduces. A descriptive length scale for the voids or meshes of the network would thus be the tube diameter $L_\perp$ as e.g. observed in fluorescence measurements (compare Figure 1.6). However, this quantity is also linked to the filaments rigidity and exhibits a non-trivial scaling behavior that we will explain below. The quantitative value of the tube diameter is a problem we address in detail in Chapter 3. A convenient measure for the mesh size ξ of the network can yet simply be defined from geometrical considerations as

$$\xi = \sqrt{\frac{3}{\nu L}} \propto c^{-1/2} \ . \tag{2.2}$$

where the polymer concentration c is proportional to the polymer contour length density νL. The mesh size's unit is length and it can be interpreted as the average distance between

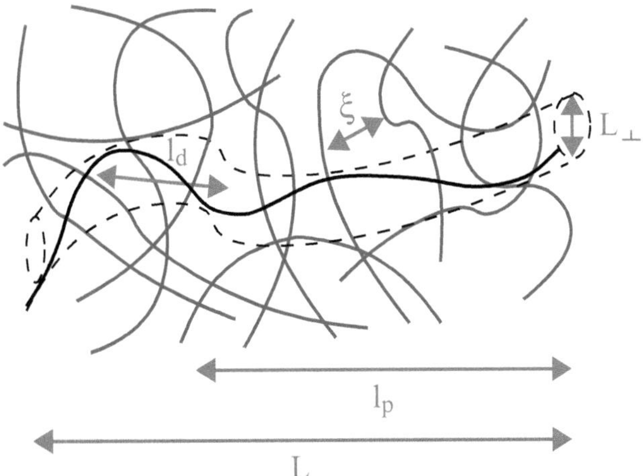

Figure 2.2: Illustration of relevant length scales in entangled networks of semiflexible polymers. The smallest length scales are the tube diameter $L_{\perp}$ and the mesh size ξ describing the properties of the network. An intermediate length scale, the deflection length l_d measures the distances between collisions of confined polymer and tube. The largest length scale describes the properties of a single polymer: contour length L and persistence length l_p.

to derive the deflection length and the tube diameter as functions of mesh size and persistence length. Our results agree to leading order with Semenov's scaling law and provide additionally an absolute value for the tube diameter and higher-order terms that account for finite length effects. Furthermore, we present the results of Monte-Carlo simulations of the system that confirm a harmonic tube potential, provide the distribution function of tube widths and reproduce the absolute tube diameter predicted by our theory.

2.2 Non-Equilibrium - Time Scales and Mechanical Response

Having derived the free energy for a network, one can also try to describe macroscopic non-equilibrium properties as for instance the mechanical response function of a polymeric material under deformation. Before we present theoretical models of viscoelasticity it is instructive to have a look at the time scales involved in the tube model. Picture a

Figure 2.4: Time scales in entangled networks: (left) below the entanglement time τ_e the confined polymer has not yet experienced interaction with the tube walls. (center) In the intermediate regime $\tau_e < t_{\text{tube}} < \tau_r$ the polymer completely explores the available tube volume. (right) At times larger than the reptation time τ_r the tube is remodeled by the sliding of the polymer along its own contour.

semiflexible polymer initially placed inside a tube potential as depicted in Figure 2.4 (left). At very short time scales this polymer will in general not interact with the surrounding filaments that constitute the tube. As long as the confined polymer has not seen or experienced the tube potential its dynamical properties like the structure factor [62, 44] or the different diffusion coefficients [63, 64] will equal the case of a free polymer. After a certain time τ_e, which is often referred to as entanglement time, the thermal fluctuations of the probe filament have resulted in encounters with the neighboring topological constraints and the complete accessible space of the tube potential has been explored as in Figure 2.4 (center). This naturally entails distinctive changes of the test filaments properties compared to free polymers. The most demonstrative of these observables is probably a sudden saturation of transversal and rotational mean square displacement describing the transient localization to the confinement tube. For much longer times however, center-of-mass diffusion of the polymer is possible by reptation of the polymer along its own contour as shown in Figure 2.4 (right). By this process, the confined filament slowly shifts into a new environment of neighbors and thus into a new segment of tube and consequently on the long run rotational diffusion still remains possible. The associated time scale is usually denoted reptation time τ_r and can be estimated as the time a filament needs to completely reptate out of its initial tube. For conventional F-actin preparations these two time scales can be estimated from experiments. While dynamic light scattering reveals the onset of restricted dynamics on the millisecond scale [65], the reptation time for long filaments can amount to several days [66, 67]. Obviously, a single filament description in terms of a fixed tube potential is thus a valid approximation on a rather extended time scale

$$\tau_e < t_{\text{tube}} < \tau_r \tag{2.7}$$

as applicable to most experimentally and biologically relevant processes. However, we will show in Chapter 4 that the tube model must not be confused with an equilibrium concept and that important consequences arise even from small scale dynamic modifications of the confinement tube. A prominent manifestation of this fact can be seen in the distribution function of local polymer curvatures. While it is usually assumed for systems of negligible excluded volume that the ensemble of confined polymers qualitatively shows the same conformation statistics as free polymers, we challenge this assumption. To this end we develop a simulation algorithm that particularly ensures the reproduction of real polymer dynamics by allowing for an effective exploration of network void spaces by breathing and short scale reptation. The validity of our simulations is confirmed by sound agreement with recent experimental measurements [57]. The observed bending distributions feature an unexpectedly high weight on highly bend filaments that can be traced back to transient entropic trapping in network void spaces. These transient non-equilibrium distributions are shown to be a generic consequence of averaging in entangled networks on intermediate time scales as relevant to experiments and most biological processes.

The time scales discussed above are also reflected in the frequency dependent mechanical response of entangled networks. The mechanical properties of polymeric materials are described by different compliances or moduli that relate strain and stress in a sample. While for general deformation fields these phenomena can only be grasped by tensors of higher dimensions in a complex mathematical formalism, idealized setups allow for a description by a single scalar - the shear modulus G. If an isotropic sample of homogeneous material is sheared as depicted in Figure 2.5 the shear modulus is defined as the ratio of stress σ to strain Γ as

$$G = \frac{\sigma}{\Gamma} = \frac{F/A}{\Delta x / L_0} \tag{2.8}$$

where the stress is the force F per cross section area A and the strain a dimensionless measure for the deformation. A different interpretation of the modulus can be inferred from the energy change associated with a shear deformation. Obtaining the potential energy by integration over the force in Eq. (2.8) results in $U = \int G A \Delta x / L_0 \mathrm{dx}$ and thereby the energy per unit volume can be expressed as

$$\frac{U(\Gamma)}{V} = \frac{U(\Gamma)}{AL_0} = \frac{G\Delta x^2}{2L_0^2} = \frac{G}{2}\Gamma^2 \,. \tag{2.9}$$

Analogous to a spring constant, the shear modulus can thus be seen as the second derivative of the deformation energy with deformation strength in the regime of linear response. This fact permits to calculate shear moduli from free energy changes and will be exploited in

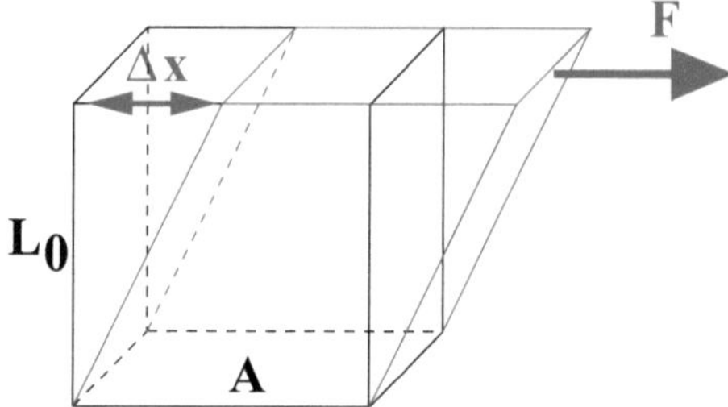

Figure 2.5: Simple shear applied to a cubic sample of height L_0 and cross-section A. The shear deformation with strain $\Delta x/L_0$ is resisted by a force F.

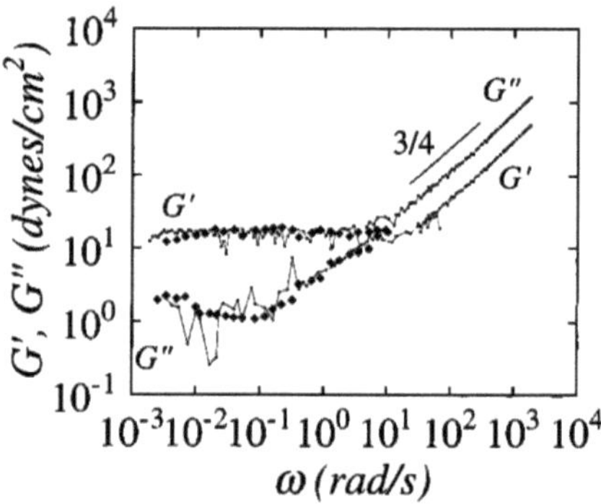

Figure 2.6: Storage modulus G' and loss modulus G'' of a $24\mu M$ F-actin solution measured as a function of shear frequency by mechanical rheometry at strains of $\Gamma = 0.01$. Reproduced from [68].

Chapter 5. For polymeric materials the response function is complicated by the fact that relaxation of stresses in the network occurs on the different time scales introduced above. Consequently, the system exhibits frequency dependent viscoelastic moduli as shown in Figure 2.6. The elastic response is given by the storage modulus G' given by the compliance that is in phase with the oscillatory probe strain and the viscous response is described by the off-phase loss modulus G''. The exact form of the two moduli and the extent of the different regimes is still subject of an ongoing discussion [69, 70, 71], but a few distinct features emerge uncontroversially: at low frequencies the response of the network obviously decreases rapidly (not shown in Figure 2.6). This is easily explained by the fact that the possibility of large scale rearrangement of the polymers by reptation permits to relax any imposed stress if the available time is sufficient, i.e. on the time scale τ_r. For high frequencies on the other hand, the system is still in the regime below the entanglement time τ_e where the polymers do not interact with their neighbors. The modulus in this regime is believed to arise mainly from the pulling out of undulations of single polymers. Theoretical models predict a scaling with frequency as $\omega^{3/4}$ [72, 73] and are in accordance with different experimental measurements [74, 75, 76, 68]. Finally, at intermediate time scales $\tau_e < t_{\text{tube}} < \tau_r$ the system can be seen as an ensemble of confinement tubes and its response can be determined from their behavior under shear. As shown in Figure 2.6 the experimentally measured moduli on this time scale stay rather constant: the elastic modulus forming a plateau and the viscous modulus featuring a dip. The response at intermediate time scales is conventionally described by the dominant elastic modulus G' at that point where the ratio G'/G'' gets maximal. This value is referred to as the plateau modulus G and will be the subject of intensive study in Chapter 5. The scaling of the plateau modulus with concentration was investigated both experimentally and theoretically. Even if the accuracy of the experimental measurements is often compromised by difficulties during sample preparation [23, 77, 78], the acquired data seems to approximately agree with a predicted scaling of the modulus with concentration as

$$G \propto c^{7/5} . \tag{2.10}$$

This scaling law originally proposed by Isambert and Maggs [79] is also obtained by conceptually different lines of argumentation [50, 80]. Isambert and Maggs build their theory on the tube model and the scaling laws by Odijk and Semenov presented above. They argue that the macroscopic deformation of the material results in corresponding deformation of the confinement tubes. Consequently, the local tube diameter increases or decreases depending on the orientation of the tube to the deformation field. Using now Odijk's picture of the deflection length it is obvious that e.g. a decrease of the tube diameter entails a decrease in the deflection length and a higher number of collision points. Since we know that every collision adds one $k_{\text{B}}T$ to the free energy and furthermore have determined the scaling of the deflection length with tube diameter Eq. (2.4) and the scaling of the tube diameter with concentration Eq. (2.6) it is easy to infer

$$G = c^{7/5} l_p^{-1/5} . \tag{2.11}$$

This scaling of the plateau modulus with persistence length and concentration will be corroborated and quantified by a detailed numerical analysis in Chapter 5.

2.3 Non-Affine Deformations

Common to all theoretical models for the plateau modulus presented above is the assumption that the deformation field is affine on all length scales. This assumption is intended to simplify the central challenge of all elastic theories of heterogeneous, composite materials: how is a macroscopic deformation passed down to the microscopic constituents? Since this is obviously a highly complex problem in a network of semiflexible polymers, it is instructive to take a look at the one-dimensional toy model depicted in Figure 2.7. Here the "material" is represented by a series of beads connected by microscopic springs. We

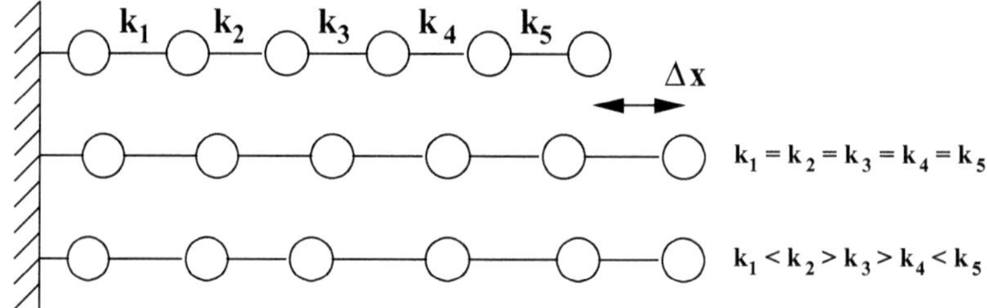

Figure 2.7: A linear chain of beads connected by springs with stiffness k_i in equilibrium (top). If the system is globally deformed by Δx, the local deformation depends on the local interactions. When all springs have equal stiffness these are homogeneous (center) and the local deformation field follows the global deformation affinely. If however the system is heterogeneous (bottom) the state resulting from an affine deformation would not be the state of lowest energy. The minimization of energy is only obtained for a non-affine local deformation.

now apply a macroscopic deformation Δx at one end of the spring chain and ask again the question how this deformation is passed down to the beads on the microscopic level. The answer is of course, that the microscopic deformation field is obtained as the set of bead positions that minimizes the bending energy of the springs. If all springs have the same spring constant $k_i = k_j$ the material can be seen as homogeneous and every bead will be deformed affinely with the macroscopic deformation. If the material however is heterogeneous and $k_i \neq k_j$, the local deformations are non-affine. While in this simple one-dimensional case the deformation field can easily be found, the task is considerably harder for a complex network of entangled polymers. Since the system is undergoing thermal fluctuations, the problem to solve is a minimization of free energy in an immense number of degrees of freedom and consequently existing theoretical approaches have resorted to the assumption of affine displacement. Obviously however, this assumption is not correctly describing the deformation field because the interactions in the network are clearly of a heterogeneous nature. In Chapter 5 we present a numerical solution to the problem and show that indeed the affine assumption only qualitatively describes the mechanical response of entangled networks while it considerably overestimates the material's stiffness. To this end we have developed an approach that permits to numerically calculate the partition sum and free energy of a polymer network. This is achieved by analyzing the free energy of one component of the transversal fluctuations of a test polymer in a two-dimensional reference frame with fluctuating point obstacles. Averaging over disorder and all possible reference frames results in an approximation for the system's free energy. Upon shear deformation we apply an algorithm that minimizes this global free energy by allowing the tube contour of the confined polymer to deviate from affine displacement. Indeed, the macroscopic plateau modulus obtained in this way agrees well with experimental data. Our results thus prove that shear deformation of networks of entangled polymers has to be correctly described in a non-affine picture and that the assumption of affine displacement leads to an overestimate of the material's response.

end, the polymer is modeled by a sequence of independent rods in Section 3.3. Criteria for the correct choice of the independent rod length and a self-consistent determination of the tube diameter are developed, before a final result for the tube diameter is obtained in Section 3.4. Extensive numerical simulations supporting these result and providing additional insight are presented in Section 3.5 followed by our conclusion in Section 3.6.

3.2 Model Definition

We consider a network of physically entangled polymers with a particular focus on pure solutions of the biopolymer F-actin. Of course, the developed theoretical framework is also applicable to all biological or synthetical polymer networks in a comparable range of parameters where strong confinement [83] into a tube is guaranteed. The polymer density ν is given by the number of polymers of length L per unit volume. The polymers are of bending stiffness κ corresponding to a persistence length $l_\mathrm{p} = \kappa/k_\mathrm{B}T$. A single polymer's configuration $\mathrm{r}(s)$ is parameterized by the arc length s and the average distance between the polymer chains can be characterized by a mesh size $\xi := \sqrt{3/(\nu L)}$ [1]. We will describe the constituent polymers by the worm-like chain model [24, 25] and exploit the tube model concept [54, 55] to reduce the description of the network to a single polymer and its neighbors. In the following we will begin our analysis with an investigation of the different length scales involved in the system.

3.2.1 Length Scales

Typical F-action solutions are polydisperse with a mean filament length $L \approx 22\mu\mathrm{m}$ [84]. With a persistence length $l_\mathrm{p} \approx 17\mu\mathrm{m}$ [22, 85] comparable to its length, it is the textbook example of a semiflexible polymer. At a concentration of $c = 0.5$ mg/ml corresponding to $\nu \approx 1\mu m^{-3}$ [86] the average mesh size equals $\xi \approx 0.4\mu\mathrm{m}$. We can thus state that the persistence length of a filament is much larger than the distance to its neighbors, $l_\mathrm{p} \gg \xi$. Since the tube diameter $L_\perp$ is at most of the order of the mesh size, this additionally implies $l_\mathrm{p} \gg L_\perp$. The polymer will thus not deviate far from the tube center. Consequently, configurations where the polymer folds back onto itself are rendered unlikely. This is a minimal requirement to model the tube by a harmonic potential of strength γ. The potential has to be seen as a hypothetical tube representing the joint contribution of all surrounding polymers which constrain the transverse undulations of a given polymer.

The energy of a certain polymer contour $\mathrm{r}(s)$ is the sum of the bending energy of the polymer and its confinement into the harmonic potential and is given in the weakly-bending rod approximation by

$$H(\gamma, \kappa) = \int_0^L ds \left[\frac{\kappa}{2}(\mathrm{r}''_\perp(s))^2 + \frac{\gamma}{2}\mathrm{r}^2_\perp(s)\right] . \tag{3.1}$$

[1]The mesh size has the unit length and can be interpreted as an average distance between network constituents. While the denominator $\sqrt{1/\nu L}$ ensures the correct scaling the numerator is a mere definition.

Here $\mathrm{r}(s) = (s, \mathrm{r}_\perp(s))$ is a parameterization in arc-length s and transverse displacement $\mathrm{r}_\perp(s) = (y(s), z(s))$ from the tube center. The prime denotes a derivative with respect to s. This harmonic approximation to the Hamiltonian of the worm-like chain model is valid as long as $|\mathrm{r}''_\perp| \ll 1$, i.e. as long as the transverse coordinates of the tube coordinate can be considered to remain single valued.

With the thermal average $\langle\cdot\rangle$ the tube diameter can now be defined as

$$L_\perp^2 := \frac{1}{2L} \Big\langle \int_0^L ds\, \mathrm{r}_\perp^2(s) \Big\rangle \, . \tag{3.2}$$

By choosing a factor of 1/2, the defined diameter is equivalent to the quadratic transverse displacement of one cartesian component only. This corresponds e.g. to the experimentally observed tube diameter in the focal plane of a microscope.

So far we have identified two length scales: the length scale of persistence length and the total polymer length describing the properties of one specific polymer, and the length scale of mesh size and the tube diameter describing the properties of the network structure. Additionally we introduce the deflection length $L_\mathrm{d} := (\kappa/\gamma)^{1/4}$ as a third useful length scale. It is interpreted below as that length on which interactions between single polymer and network occur. More precisely, it is a measure for the number of contacts of the polymer with the tube walls. For large confinement strength γ the tube is small, making interaction with the encaged polymer more likely and therefore resulting in a small deflection length. On the other hand, for a large polymer rigidity κ transverse undulations allowing contacts with the tube walls are energetically unfavorable and the distance between contact will decrease. For $l_\mathrm{p} \gg L_\perp$ we expect the deflection length to be distinctively smaller than the polymer length, but also larger than the tube diameter. For quantification we consider the free energy cost $\Delta F(\gamma)$ of confining the polymer to the tube. It can be found from the partition sum that is obtained as a path integral over all polymer configurations:

$$\exp\left[-\beta \Delta F(\gamma)\right] = \int \mathcal{D}[\mathrm{r}_\perp(s)] \exp[-\beta H(\kappa, \gamma)] \tag{3.3}$$

with $\beta = 1/k_\mathrm{B}T$. In the limit of infinitely long polymers the free energy cost is [59]

$$\Delta F = \sqrt{2} k_\mathrm{B} T \, \frac{L}{L_\mathrm{d}} \, . \tag{3.4}$$

This result fits into the picture of the deflection length as measure for the average distance between successive collisions of the polymer with its tube. If the typical distance between two collisions is given by L_d, the free energy loss results as the sum over all L/L_d points of contact where every collision costs one $k_\mathrm{B}T$. The free energy now allows one to derive the tube diameter as

$$L_\perp^2 = \frac{1}{L} \frac{\partial \Delta F}{\partial \gamma} = \frac{L_\mathrm{d}^3}{2\sqrt{2} l_\mathrm{p}} \, . \tag{3.5}$$

In the limit of infinite polymer length we have thus derived the tube diameter as a function of the deflection length by differentiation of the free energy cost.

The above consideration also sets the road map for the remaining work. To calculate the tube diameter for the network, we need first to connect free energy and tube diameter for polymers of finite length and then derive the deflection length for the model under investigation.

3.2.2 Finite length Polymers

For finite size polymers the path integral in Eq.(3.3) can be evaluated exactly [59, 87] and with the dimensionless deflection length $l_\mathrm{d} := L_\mathrm{d}/L$ results in

$$\Delta F = -2k_\mathrm{B}T g(l_\mathrm{d}) \tag{3.6}$$

with

$$g(l_\mathrm{d}) = \ln(l_\mathrm{d}^2) - \frac{1}{2}\ln\left(\sinh^2\frac{1}{\sqrt{2}l_\mathrm{d}} - \sin^2\frac{1}{\sqrt{2}l_\mathrm{d}}\right) . \tag{3.7}$$

The limit of small l_d that is guaranteed by $L \gg L_\mathrm{d}$ as stated above, allows an expansion

$$g(l_\mathrm{d}) = -\frac{1}{\sqrt{2}l_\mathrm{d}} + \ln(l_\mathrm{d}^2) + \mathcal{O}(e^{-1/l_\mathrm{d}}) \quad \text{for} \quad l_\mathrm{d} \to 0 , \tag{3.8}$$

where the first term is just the result for polymers with infinite length (3.4). Upon again using the relation $L_\perp^2 = (1/L)(\partial\Delta F(l_\mathrm{d})/\partial\gamma)$ with the inner derivative $\partial l_\mathrm{d}/\partial\gamma = -(L^4/4\kappa)l_\mathrm{d}^5$ the tube diameter becomes

$$L_\perp^2 = \frac{L^3}{2l_\mathrm{p}} l_\mathrm{d}^5 g'(l_\mathrm{d}) . \tag{3.9}$$

For later convenience we simplify this to $l_\perp^2 = h(l_\mathrm{d})$ by introducing a dimensionless tube width $l_\perp$ and function $h(x)$ as

$$l_\perp^2 := \frac{L_\perp^2 l_\mathrm{p}}{L^3} \qquad \text{and} \qquad h(x) := \frac{x^5 g'(x)}{2} . \tag{3.10}$$

This relation connects the wanted tube diameter to the deflection length and hence to the hypothetical tube potential γ at a given bending rigidity. In the following we will further investigate the tube properties and set up a model that allows one to derive the deflection length and thereby the hypothetical harmonic tube potential strength from the polymer concentration and persistence length.

3.3 Independent Rod Model

For simplification and as an anticipation towards the computer simulations, consider for the time being a polymer in a two-dimensional (2D) plane. In this case the transverse displacement vector $\mathrm{r}_\perp$ reduces to a single component. The undulations of the test polymer in 2D are hindered by point-like obstacles as depicted in Figure 3.1 (top). These obstacles represent the cuts of the surrounding polymers in three dimensions with the chosen fluctuation plane. Given an appropriate number of 2D obstacles equivalent to the density of surrounding polymers in 3D, the transverse displacement will correspond to one of the two components of the displacement vector $\mathrm{r}_\perp$, if we assume the fluctuations of these components to be independent. Bearing in mind the large persistence length compared to the mesh size, the surrounding polymers in 3D are modelled as rigid rods and the area density ρ_{MC} of obstacles in 2D corresponding to a polymer concentration ν in the 3D network is $\rho_{\mathrm{MC}} = 2\nu L/\pi$. It is computed in Appendix A and will be explicitly needed for the comparison with simulation results.

Recalling the Hamiltonian (3.1), the polymer's free energy has a bending and an entropic contribution. To minimize the free energy it can be favorable to trade in bending energy for a wider tube. Thereby entropy is gained due to a larger available free volume, but the polymer is forced to sacrifice energy to obtain its curvature (see Figure 3.1 (top)). This competition defines a characteristic length $\bar{L}$ that has to be of the order of the deflection length L_{d}, since this is the length scale characterizing interaction of the test polymer and its environment.

In the following we will develop an analytical theory based on an independent rod model (IRM) that is inspired by the competition we have just discussed. To this end, we use a simplified model of a semiflexible polymer, in which the flexibility is localized to the joints of a sequence of independent stiff rods of length $\bar{L}$. After deriving the transverse fluctuations of a single independent rod in an environment of fluctuating neighbors, we apply a self-consistency argument to arrive at the corresponding tube width of the full length semiflexible polymer. Note that the analysis is carried out for three dimensions and the 2D simplification only serves for illustration and for simulations later on.

To begin with, consider the test polymer to be divided into independent segments of length $\bar{L}$ that are assumed to be completely rigid rods and are only allowed to undergo transverse fluctuations. As the flexibility in the IRM depends on the number of joints, it is obvious that the choice of $\bar{L}$ is crucial for the resulting tube diameter. Picturing the decomposition of the test polymer as in Figure 3.1 (middle) it can be seen that the transverse fluctuations of the independent rods are hindered by the two closest obstacle polymers normal to either side of each segment of length $\bar{L}$. If $\bar{L}$ is chosen too large (e.g. $\bar{L} = L$ in the worst case) the area of transverse fluctuations will be much smaller than for a true semiflexible polymer because flexibility is underestimated (Figure 3.1 (bottom, right)). On the contrary, if $\bar{L}$ is chosen too small, the normal distance to the nearest obstacle can be quite large (Figure 3.1 (bottom, left)). This overestimation of flexibility results in a transverse fluctuation area that is large compared to the polymer we try to model. Before we further discuss the proper choice of $\bar{L}$, we will focus on the behavior of a single independent rod in more detail.

The transverse fluctuation of a single stiff rod in the (y, z)-plane are constrained by the projections of the surrounding network constituents to this plane as depicted in Figure 3.2 (left). Since the mesh size is much smaller than the persistence length, the surrounding polymers can be assumed to be straight and "dangling ends" are neglected. The size of the shaded cross section will decrease with increasing density of polymers, i.e. with a decreased mesh size. Thus the tube diameter is of order of the mesh size and scales as

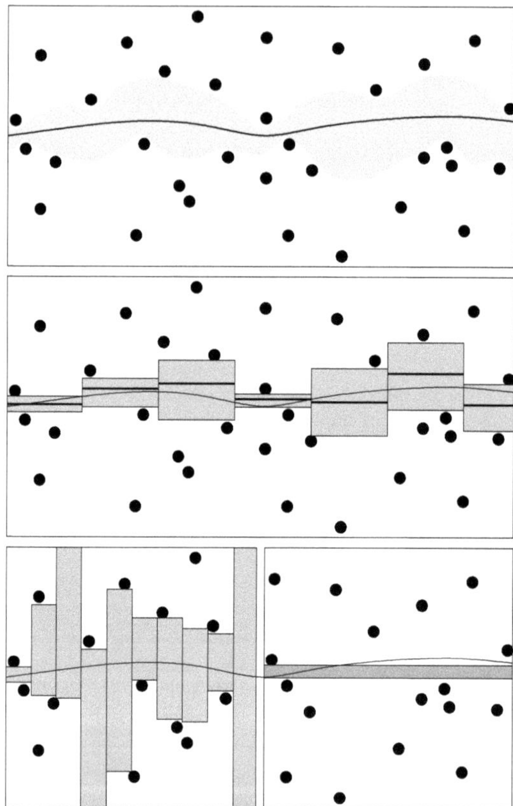

Figure 3.1: (top) The fluctuation tube of a semiflexible polymer in a network of constraints is determined by a delicate balance of entropic and bending energy. (middle) Scheme of decomposition of a semiflexible polymer into rigid rods of length $\bar{L}$. The flexibility is localized to the joints between independent rods. Given the proper choice of $\bar{L}$ both models produce the same transverse fluctuation area. (bottom) Small rod length overestimates and large rod length underestimates fluctuations.

$L_\perp \propto \xi$ for a given 2D plane. Furthermore, an increase of the length $\bar{L}$ of the rigid rod signifies an increase of obstacles that will be projected to the plane. As the average distance between surrounding polymers in direction of the test rod is also given by the mesh size ξ, the average number projected onto the plane increases as $\bar{L}/\xi$. As this reduces the cross section area, we finally arrive at an overall scaling of the tube diameter as $L_\perp \propto \xi^2/\bar{L}$.

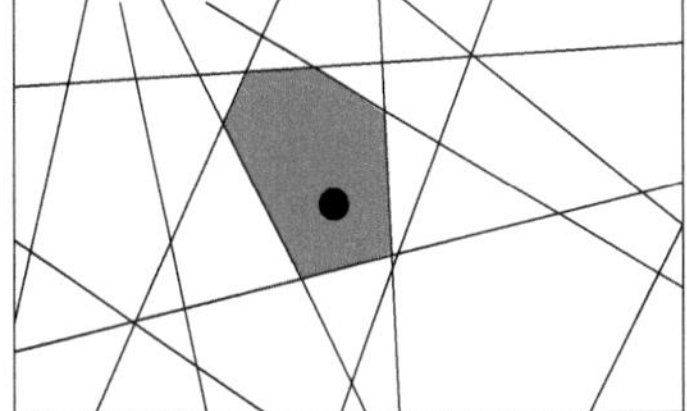
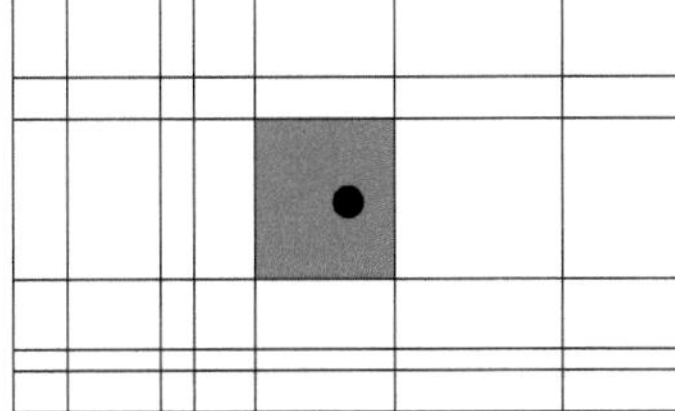

Figure 3.2: (left) Projection of constraining polymers to the plane of transverse fluctuations of a test polymer (black dot). As the mesh size is much smaller than the persistence length, the constraining filaments, can be assumed to be straight. The shaded area is the accessible tube area for a specific obstacle configuration. (right) Corresponding setup for a simplified geometry where obstacles can only be aligned with coordinate axes.

Before we quantify this scaling result in the next section, let us first have a closer look at the obstacles. In a self-consistent treatment these evidently have themselves to be regarded as semiflexible polymers of the network and therefore undergo fluctuations around an average position as well. This causes the cross section area to smear out, as the test polymer has now a non-vanishing probability to take on values behind the average obstacle position. In terms of a confinement potential the cross section is no more described by an infinite well, but by some continuous potential which earlier has been assumed to be harmonic with strength γ per unit length of a polymer. The obstacle fluctuations will also be modelled as Gaussian and to distinguish between the test polymer mean square displacement $L_\perp^2$ and the obstacle's, the latter is denoted as σ^2. In a self-consistent treatment of the network the average tube width $L_\perp$ of the test polymer is then determined as a function of the obstacle fluctuations σ, where σ is chosen such that $L_\perp = \sigma$. Of course, the value $L_\perp$ of a single obstacle configuration will not only depend on σ but also on the obstacle positions in that specific configuration. Consequently, averaging over all obstacle configurations will result in a distribution $P(L_\perp)$ and self-consistency would then also require a distribution $P(\sigma)$. However, if we assume these distributions to be reasonably peaked, we can use their averages as a good approximation. The self-consistency of distributed tube widths is verified by simulations in Section 3.5.

3.3.1 Single stiff rod in simplified geometry

According to the assumptions made above the obstacles (in a top view) are completely described by a normal distance r_k from the test polymer and an orientation α_k; compare Figure 3.2. We will neglect correlations and assume the obstacles to be uniformly distributed. The probability to find an obstacle with a certain direction at a specified point is independent of the direction and that point. This corresponds to a complete factorization of the network distribution function into single polymer distribution functions.

Consider first a simplified geometry in which all obstacles are either parallel to the y or the z axis as depicted in Figure 3.2 (right). As fluctuations in both coordinates are assumed to be independent and equivalent, the task of computing the tube width is reduced to a one dimensional problem with a single coordinate r. The network density or mesh size enters as the number of obstacles per unit length ρ. This density should be chosen such, that the average number of obstacles at a certain distance r from the test rod in the IRM is the same as the average number of obstacle polymers featuring a minimal distance r from the test polymer. This density is proportional to the length $\bar{L}$ of the stiff segment and the number of surrounding polymers in a unit volume νL. The exact relation $\rho = (\pi/2)\nu L\bar{L}$ is calculated in Appendix B. Note that in the simplified geometry the obstacle density is not a complete radial density but a line density of obstacles on one of the four axes (positive and negative y and z axis). To recover the complete radial density, one has to sum over all of these. Hence the obstacle density on either one of the four axes has to be $\rho/4$.

As the obstacles are assumed to undergo Gaussian fluctuations around their average position r_k, the corresponding probability density is

$$P_0(r - r_k, \sigma) := \left(2\pi\sigma^2\right)^{-1/2} e^{\frac{-(r-r_k)^2}{2\sigma^2}} . \tag{3.11}$$

If the test rod interacts with only a single obstacle, we can state that the probability to find the test rod at a certain position is given by the fraction of realizations still accessible to the obstacle. In this case

$$P_+(r, r_k, \sigma) = \int_r^\infty dr' P_0(r' - r_k, \sigma) \tag{3.12}$$

is the fraction of configuration space still accessible to the obstacle if the test rod is placed at r (for $r_k > 0$). Completing the integral yields

$$P_+(r, r_k, \sigma) = \frac{1}{2}\mathrm{erfc}\left(\frac{r - r_k}{\sqrt{2}\sigma}\right) \tag{3.13}$$

and the corresponding probability for obstacles at negative positions $P_-(r, r_k, \sigma)$ is simply obtained by a inverted sign of the argument. The probability to find the test rod at a position r for a given configuration of obstacles $\{r_k\}$ is then given by the product of all probabilities

$$P(r, \{r_k\}, \sigma) = \frac{1}{\mathrm{N}} \prod_{k, r_k > 0} P_+(r, r_k, \sigma) \prod_{k, r_k < 0} P_-(r, r_k, \sigma) . \tag{3.14}$$

The normalization $\mathrm{N} = \mathrm{N}(\{r_k\}, \sigma)$ is determined by the condition $\int drP(r, \{r_k\}, \sigma) = 1$ and depends on the obstacle configuration.

As the function $P_+(r, r_k, \sigma)$ reduces to a Heaviside function in the case of $\sigma \to 0$, the product in Eq. (3.14) can be written as $\theta(r - r_-)\theta(r - r_+)/(r_+ - r_-)$ where r_+ and r_- are the positions of the two closest obstacles. This reduction is justified because all further obstacles are completely shadowed by the two nearest neighbors. In the case of a non-vanishing σ the probability distribution $P(r, \{r_k, \alpha_k\}, \sigma)$ will not be rectangular anymore but smear out. The test rod has a non-vanishing probability to be found behind the average position of the closest obstacle and thus a chance to feel the interaction of further network constituents. However, sketching the distribution in Figure 3.3, it becomes intuitively clear that this probability rapidly approaches zero for far obstacles or small fluctuation amplitudes σ. We will exploit this fact in the numerical analysis below and in the simulations.

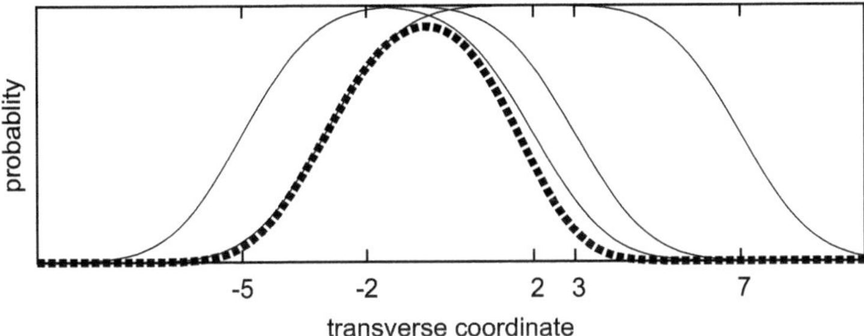

Figure 3.3: Probability density to find the test rod at a spatial position for mutual interaction with a single obstacle (solid lines) and resulting probability in an environment of all obstacles (dashed line). The x-axis tics mark the center position of each obstacle. Distant obstacle only have a negligible influence on the overall probability function.

Using the distribution function Eq. (3.14) for the test rod, averages of any function $f(r)$ can now be calculated for a single realization of obstacles as

$$\overline{f(r)}_{\{r_k\}} = \int drf(r)P(r, \{r_k\}, \sigma) \,, \tag{3.15}$$

where the index $\{r_k\}$ denotes the specific obstacle configuration. The tube center of the test rod is then

$$\overline{r}(\{r_k\}, \sigma) := \overline{r}_{\{r_k\}} \tag{3.16}$$

and the width of the probability distribution is the wanted tube diameter

$$L_\perp^2(\{r_k\},\sigma) := \overline{r^2}_{\{r_k\}} - \overline{r}^2_{\{r_k\}} \,. \tag{3.17}$$

The derived tube diameter of the test rod is not only a function of the fluctuation width σ but also of the specific obstacle configuration. Consequently, sampling over different obstacle sets will result in a distribution of values for $L_\perp$. As mentioned earlier this distribution should be described by a single characteristic value - consistent with the obstacle fluctuations that have also be assumed to be of equal size. Since the obstacles are uniformly distributed, they can be fully described by the density encoded in the average number of obstacles per line ρ. This is achieved by integrating out all obstacle positions and orientations in $L_\perp^2(\{r_k\},\sigma)$ to arrive at the only density dependent $L_\perp^2(\rho,\sigma)$. We choose a simple average over a large number N of obstacle sets $\{r_k\}$ like

$$\langle f(\{r_k\})\rangle_\rho = \left(\prod_{k=1}^{N}\int_{-R/2}^{R/2}\frac{dr_k}{R}\right) f(\{r_k\}) \,, \tag{3.18}$$

where $R = N/(\rho/4)$. In this nomenclature the average tube diameter [2] is obtained as $L_\perp^2(\rho,\sigma) = \langle L_\perp^2(\{r_k\},\sigma)\rangle_\rho$.

Self-consistency is now expressed as

$$L_\perp^2(\rho,\sigma) = \sigma^2 \tag{3.19}$$

at the point of self-consistency (PSC) $\sigma = \sigma^*$. By measuring length in $1/\rho$ we can rewrite this to a dimensionless master curve $l(\rho\sigma)$:

$$L_\perp^2(\rho,\sigma) = \frac{1}{\rho^2} l(\rho\sigma) \,, \tag{3.20}$$

since $L_\perp, 1/\rho$ and σ are all lengths, The task of finding the self-consistent tube width $L_\perp(\rho,\sigma^*) = \sigma^*$ translates to finding $l(C) = C^2$ where the constant $C = \rho\sigma^*$. As soon as this is achieved, the self-consistent tube diameter is available as a function of density ρ only and hence it depends like

$$L_\perp = \frac{C}{\rho} = \frac{2C}{3\pi}\frac{\xi^2}{\bar{L}} \tag{3.21}$$

on the rod length $\bar{L}$ and mesh size ξ.

The numerical determination of C is achieved by an integration using a Monte-Carlo procedure. It includes the N-fold integrals over the obstacle positions r_k from Eq. (3.14) as well as the integration over the test polymer position r from Eq. (3.15). As mentioned above, the probability distribution rapidly decreases at distances far from the closest obstacle. Hence, we have restricted the integration range of the integration over r in the

[2]Of course, one could also image a different characterization of the average tube diameter, e.g. the median or the maximal diameter. We choose the average as the most obvious quantity experimental groups might measure, e.g. in analyzing different fluorescent microscopy images.

Monte-Carlo samples to values $[y_{min} - 5\sigma, y_{max} + 5\sigma]$ where y_{min} and y_{max} are the closest obstacle at either side. Furthermore, the fast decrease of the probability distribution renders the contribution of distant obstacles quasi to zero. We can therefore drop all obstacles with $y_k \notin [y_{min} - 10\sigma, y_{max} + 10\sigma]$. The results are depicted in Figure 3.4 and a graphical solution for the PSC constant results in

$$C \approx 3.64 \,. \tag{3.22}$$

Special attention should be paid to the behavior at $\rho\sigma = 0$. It provides a good test whether the used IRM is adequate and allows for a verification of the numerics. At a finite density as required by the tube concept, $l(0)$ reflects the situation of immobile obstacles with $\sigma = 0$. At this point the tube diameter should remain finite and its value should be given by the density of obstacles. From the obstacle statistics and density per unit length $\rho/4$, the probability to find the first obstacle at position $r_\pm$ is known to be $P(r_\pm) = \exp(-r_\pm\rho/4)$. In the case of fixed obstacles the available fluctuation area is $2L_\perp = r_+ - r_-$ and the expectation value $4\langle L_\perp^2\rangle = \langle (r_+ - r_-)^2\rangle$ can be computed from the probability density above. Taking care of the normalization one arrives at $L_\perp = \sqrt{8}/\rho$. The master function yields $l(0) = \rho^2 L_\perp^2 = 8$, a value in good agreement with the data (circles) in Figure 3.4.

3.3.2 Generic 2d Geometry

If the simplification of axis-parallel obstacle polymers is dropped again, the obstacle configuration needs to be specified by a set of radii $\{r_k\}$ and angles $\{\alpha_k\}$. The probability to find the test rod at a position (y, z) for a given configuration of obstacles $\{r_k, \alpha_k\}$ in the two-dimensional case as in Figure 3.2 (left) is then again given by the product of all probabilities where different angles have to be accounted for:

$$P(y, z, \{r_k, \alpha_k\}, \sigma) = \frac{1}{\mathrm{N}} \prod_k P_\pm(y\cos\alpha_k + z\sin\alpha_k, r_k, \sigma) \,. \tag{3.23}$$

The normalization factor $\mathrm{N} = \mathrm{N}(\{r_k, \alpha_k\}, \sigma)$ is again determined by the condition

$$\int \mathrm{dy} \quad \mathrm{dz} P(y, z, \{r_k, \alpha_k\}, \sigma) = 1 \,. \tag{3.24}$$

In a single obstacle configuration the tube diameters $L_{\perp y,z}$ in the y and z direction will in general be different. However, in averaging over all configurations isotropy must be recovered to show

$$L_\perp^2(\rho, \sigma) = \langle L_{\perp y}^2(\{r_k, \alpha_k\}, \sigma)\rangle_\rho = \langle L_{\perp z}^2(\{r_k, \alpha_k\}, \sigma)\rangle_\rho \,. \tag{3.25}$$

The average over obstacle configurations at fixed density of uniformly distributed obstacles is performed as

$$\langle f(\{r_k, \alpha_k\})\rangle_\rho = \left(\prod_{k=1}^{N} \int_0^R \frac{dr_k}{R} \int_0^{2\pi} \frac{d\alpha_k}{2\pi}\right) f(\{r_k, \alpha_k\}) \tag{3.26}$$

with integration range being again $R = N/\rho$. Note that contrary to the simplified geometry the obstacle density per unit length ρ in this case is given by $\rho = (\pi/2)(\nu L\bar{L})$. Evaluating the integrals in Eq. (3.26) again by the Monte-Carlo method results in the data plotted in Figure 3.4 (triangles), where suppression of irrelevant obstacles was implied analog to the simplified geometry. The results do not deviate much from the data obtained earlier

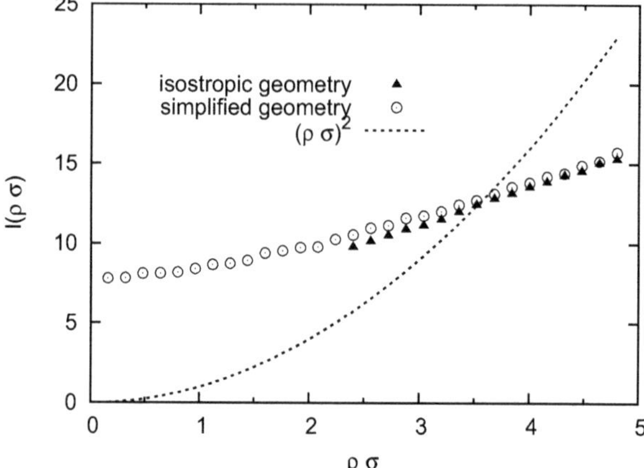

Figure 3.4: Master curve $l(\rho\sigma)$ of the tube diameter rescaled by obstacle density obtained by MC simulation for simplified (circles) and generic geometry (triangles); intersection with quadratic obstacle fluctuation amplitude marks the point of self-consistence. The error of the simplified geometry is surprisingly small.

(circles), i.e. the mistake in using a simplified geometry is surprisingly small. Again the value of the PSC is obtained graphically. It yields

$$C \approx 3.52 \tag{3.27}$$

and will be used in the remainder of this work.

3.3.3 Choice of Independent Rod Length

As discussed before and illustrated in Figure 3.1 (bottom) the choice of $\bar{L}$ is crucial for the success of the IRM. The number $L/\bar{L}$ of independent rods can be regarded as a measure for the flexibility of the modeled polymer and has to be chosen such that the transverse excursions of the ensemble of stiff rods equal the fluctuations of the actual semiflexible polymer. To this end we consider both systems in a generic harmonic potential

$$U[y(s)] = \frac{\gamma}{2}\left[y(s) - y^0(s)\right]^2 \tag{3.28}$$

with the potential minimum $y^0(s)$ as a Gaussian variable with $\langle y^0(s)y^0(s')\rangle = \alpha\delta(s-s')$. This corresponds to the assumption of a "Gaussian random backbone" as a general property of the tube. We use this intuitive assumption as one possible prerequisite to determine the segment length $\bar{L}$. Of course, other possibilities can be imagined. Note that the simulations in Sec. 3.5 will justify this assumptions a posteriori.

The average position $\overline{y(s)}$ as a functional of a given potential $y^0(s)$ is obtained as an average over all polymer configurations in this potential. Averaging then over all potential conformations yields the the mean square of the polymer's transverse fluctuations $\langle\overline{y(s)}^2\rangle$. The over-line thus denotes an average in a given potential and the brackets denote an average over all potentials. While the transverse fluctuations of a rigid rod are a function of the potential parameters α, γ only, the response of a semiflexible polymer will additionally depend on its stiffness. This evidently provides a tool to connect the semiflexible polymer persistence length and the length $\bar{L}$ from the IRM by demanding that the fluctuations $\langle\overline{y(s)}^2\rangle$ for given potential parameters α, γ are the same for both cases.

Starting with the IRM, it is sufficient to consider only one stiff rod, as the individual rods are statistically independent. The average position is then

$$\overline{y} = \frac{1}{\bar{L}}\int_0^{\bar{L}} ds\; y^0(s) \tag{3.29}$$

and the transverse fluctuations

$$\langle\overline{y}^2\rangle = \frac{1}{\bar{L}^2}\int_0^{\bar{L}} ds \int_0^{\bar{L}} ds' \langle y^0(s)y^0(s')\rangle = \frac{\alpha}{\bar{L}}\,. \tag{3.30}$$

For the semiflexible polymer the fluctuations of polymer and tube potential are decomposed into modes (Appendix C):

$$\langle\overline{y}^2\rangle = \frac{1}{L}\int_0^{L} ds\langle\overline{y(s)}^2\rangle = \frac{1}{L}\sum_k\langle\overline{y_k}^2\rangle\,, \tag{3.31}$$

where the mode analysis yields $\overline{y_k} = y_k^0/(1+q_k^4 l_{\rm d}^4)$ (compare Eq. (C.3)) with $q_k \approx \pi(k+1/2)$. Using now the correlations of the Gaussian random tube profile and the identity (C.5) the polymer fluctuations can be related to the deflection length as:

$$\langle\overline{y}^2\rangle = \frac{1}{L}\sum_k \frac{\langle(y_k^0)^2\rangle}{(1+q_k^4 l_{\rm d}^4)^2} = \frac{\alpha}{L}\frac{h'(l_{\rm d})}{4 l_{\rm d}^3}\,. \tag{3.32}$$

Equating the fluctuations for the IRM and the semiflexible polymer fixes the segment length to

$$\bar{L} = L\frac{4 l_{\rm d}^3}{h'(l_{\rm d})}\,. \tag{3.33}$$

Concluding the last section, we have obtained the tube diameter for a sequence of independent rods of length $\bar{L}$ and derived a condition how to fix this length to correctly mimic the behavior of a semiflexible polymer in a network of same mesh size. It has turned out the the criteria for the correct rod size is a function of the deflection length.

3.4 Results

If we recall that the tube diameter for a semiflexible polymer was derived in Sec. 3.2 from the Hamiltonian with a likewise dependence on deflection length, we are now equipped to set up an implicit equation to determine this deflection length. Afterwards the tube diameter can be derived from a simple calculation.

Equating the expressions for the tube diameter of the polymer (3.10) and the IRM (3.21) respectively yields

$$L_\perp^2 = \frac{L^3}{l_\mathrm{p}} h(l_\mathrm{d}) = \frac{4C^2}{9\pi^2} \frac{\xi^4}{\bar{L}^2} \, . \tag{3.34}$$

With the correct rod length (3.33) the implicit equation for the dimensionless deflection length is

$$h(l_\mathrm{d}) = \frac{C^2}{36\pi^2} \frac{[h'(l_\mathrm{d})]^2}{l_\mathrm{d}^6} \frac{l_\mathrm{p}\xi^4}{L^5} \, . \tag{3.35}$$

Solving this equation, determines l_d from the system's parameter l_p, L and ξ. It is achieved by introducing a dimensionless function

$$l_\mathrm{p}\xi^4/L^5 = j(l_\mathrm{d}) := \frac{l_\mathrm{d}^6 36\pi^2 h(l_\mathrm{d})}{C^2[h'(l_\mathrm{d})]^2} \, . \tag{3.36}$$

Inversion then yields

$$l_\mathrm{d} = j^{-1}(l_\mathrm{p}\xi^4/L^5) \, . \tag{3.37}$$

And we finally obtain for the dimensional deflection length to first order and second order in $l_\mathrm{p}^{1/5}\xi^{4/5}/L$:

$$L_\mathrm{d} = \frac{C^{2/5}}{2^{7/10}\pi^{2/5}} \xi^{4/5} l_\mathrm{p}^{1/5} + \frac{C^{4/5} 2^{1/10}}{3\pi^{4/5}} \frac{\xi^{8/5} l_\mathrm{p}^{2/5}}{L} \, . \tag{3.38}$$

By application of (3.10), the tube diameter is easily obtained as

$$L_\perp = \frac{C^{3/5}}{2^{18/10}\pi^{3/5}} \frac{\xi^{6/5}}{l_\mathrm{p}^{1/5}} + \frac{C}{2\pi} \frac{\xi^2}{L} \, . \tag{3.39}$$

Evaluation of the numerical factors holds the following results to first order

$$L_\mathrm{d} \approx 0.64 \xi^{4/5} l_\mathrm{p}^{1/5} \, , \qquad L_\perp \approx 0.31 \frac{\xi^{6/5}}{l_\mathrm{p}^{1/5}} \, . \tag{3.40}$$

Note that this also determines the confinement free energy of the polymer to $\Delta F \approx 2.21\, k_\mathrm{B}T \frac{L}{\xi^{4/5} l_\mathrm{p}^{1/5}}$. The leading term of the tube diameter agrees with the established scaling [60]. The additional term's dependence on the inverse polymer length indicates a finite length effect. It can be traced back to the partition sum of a finite polymer (3.8) and accounts for boundary effects at the end of the tube. If the free energy of infinite polymers (3.4) is used throughout the calculations, all higher order terms vanish accordingly. In

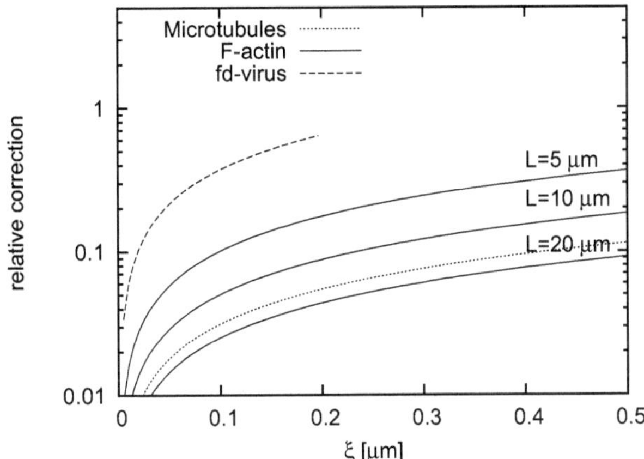

Figure 3.5: Relative correction obtained by the second order term of the tube diameter (3.39) for different biopolymers as a function of mesh size ξ. fd-viruses ($L \approx 0.9\mu$m, $l_{\mathrm{p}} \approx 2.2\mu$m [88]) show a large correction due to their small length compared to the mesh size, while this effect is rather small in microtubules ($L \approx 50\mu$m, $l_{\mathrm{p}} \approx 5000\mu$m [19, 22]). The correction for F-actin has been plotted for different length from a typical length distribution. The correction for fd-viruses has only been plotted in a range where $L_D > \xi$.

an earlier work [80] another prefactor of $L_\perp \approx 0.41\xi^{6/5} l_\mathrm{p}^{-1/5}$ for the scaling term has been predicted by rather different accounting of obstacles. For a verification of the validity of the derived absolute values we present numerical simulations below.

It is important to be aware of the subtle difference between the explicit length dependence of the first order term and the implicit dependence on L that enters via the mesh size $\xi = \sqrt{3/\nu L}$. In a polydisperse polymer solution the L in the mesh size has to be the average polymer length, while the L in the second order terms is the length of the actually observed filament in the tube. While in an idealized monodisperse solution these quantities are identical, real polymer networks feature a broad distribution of filament lengths due to polymerization dynamics. To our knowledge the derived finite length corrections provide the first theoretical description that allows to account for effects of polydispersity. Experimentally observed distributions [89, 48] are still under debate but mostly feature a pronounced slope towards lengths below the mean. Therefore, integration over the distribution with respect to the second order term will in general result in an increased average tube diameter as small lengths dominate the correction term.

The importance of the second order term depends heavily on the nature of the polymers making up the network. In Figure 3.5 the relative tube width correction obtained by the second order term is displayed for several semiflexible biopolymers as a function of mesh size ξ. It is interesting to note that the intuitive dependence on the relative persistence length l_p/L present in the second order term of the deflection length is rather negligible. The most dominant effect of the correction term is not obtained for the stiffest biopolymer, a microtubule, but for the small fd-virus. This is due to its small length to mesh size ratio. Finite length effects will influence a large fraction of the polymer strand and not only the boundaries. Given a proper control of polymer length, this effect should be experimentally observable in F-actin solutions.

Focussing back on F-actin, Figure 3.6 displays the result of our model in comparison to experimental data [48, 90]. While theoretical and experimental results are certainly qualitatively comparable, a more detailed discussion is difficult due to the large fluctuations of the measurements. However, it seems reasonable to interpret these measurements regardless of their ambiguity as an upper limit to the tube diameter. Two main reasons cause an experimental observation of tube widths systematically higher than in the presented theory: from a technical point of view the microscope resolution broadens the observed tubes. Additionally, this effect is further enhanced by collective fluctuations of the complete elastic medium that remain unaccounted for in our approach. Contrary, the computer simulations presented below, can be tailored to avoid these effects and study the exact model system used by the theory.

3.5 Simulations

We have conducted intensive numerical simulations of the model system for several reasons: on the one hand they serve as a tool to verify the validity of several approximations used in the theoretical description developed above, being for example the harmonic description

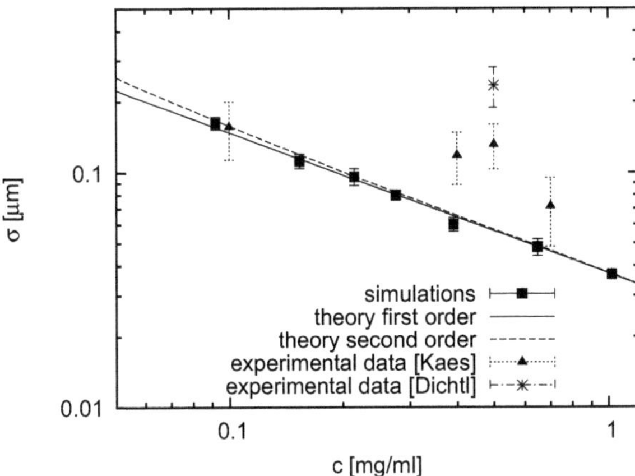

Figure 3.6: Comparison of tube diameter from theory, numerical simulations (squares) and reanalyzed experimental measurements (triangles) from [48, 90]. While Dichtl has directly measured potential strengths, Käs has recorded the maximal tube width a. Therefrom we estimated a lower boundary of $\sigma = a/6$.

of the tube potential or the assumption of a single fluctuation amplitude for the obstacles. Furthermore, the comparison between the simulated transverse fluctuations and the final result of our theory can prove if we have succeeded in correctly predicting the tube diameter in a network of semiflexible polymers. In particular, this allows to independently verify if the description in the IRM was justified. Finally, the simulations give us the chance to analyze observables that go beyond the analytical theory presented. These are in particular distribution functions and open up a further possibility to comparison with experiments.

We use a Monte Carlo simulation of a single polymer in two dimensions that is surrounded by point-like obstacles. This reduction will result in an equal fluctuation amplitude as in the 3D model, because we have assumed the fluctuations along the different coordinates to be independent. Simulating a test polymer in 2D and measuring its transverse displacement, will thus on average correspond to either $L_{\perp,y}$ or $L_{\perp,z}$ given that the number of obstacle points has been chosen correctly. We calculate this number as the number of stiff rods that cut an arbitrary unit area plane if these rods are of length L, density per unit volume ν and equally distributed both in position and orientation. The approximation as rigid rods is justified by the large persistence length compared to the mesh size. The relation between polymer concentration ν and point-density in simulations ρ_{MC} then yields (A.2):

$$\rho_{\mathrm{MC}} = \frac{2}{\pi}\nu L \ . \tag{3.41}$$

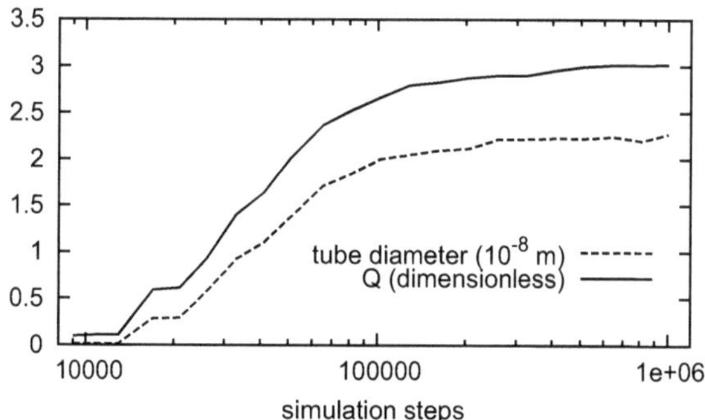

Figure 3.8: After sufficient simulation time the ratio Q (solid) approaches the characteristic value $Q = 3$ of a Gaussian distribution. The transverse fluctuation area (dashed) converges likewise.

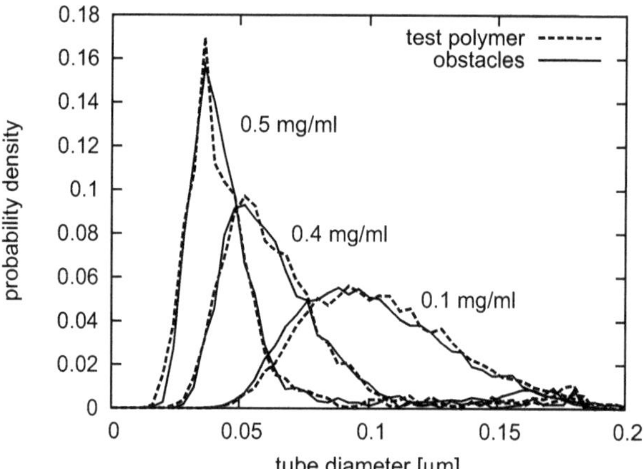

Figure 3.9: Distribution of $L_\perp$ sampled over polymer arc-length and different obstacle environments for three designated polymer concentrations in mg/ml. Distributions are well-peaked and exhibit longer tails at large tube widths. The noise at diameters far from the distributions maximum is an artefact from the numerical discretization.

consistence is guaranteed in spite of the simplifying assumption of a single fluctuation width. To this end, we have used the resulting histogram of tube diameters from above to compute a normalized distribution function. The fluctuation width of the obstacles are now initialized according to this very distribution. The resulting histogram of tube diameters is then again fed back into the simulation as obstacle fluctuation distribution. This procedure is carried out until both distributions converge against each other in a self-consistent manner. Surprisingly, this is already the case after the first iteration step of the process as displayed in Figure 3.9. This gives strong evidence that due to the self-averaging over obstacles the modelling of a network with Gaussian tube profile and a single average tube diameter is sufficient to describe the physical reality.

3.6 Conclusion

We have presented a new approach to determine the absolute value of the tube diameter in semiflexible polymer networks supported by computer simulations. To this end the deflection length of a polymer in a hypothetical harmonic tube was connected to the tube's diameter via the free energy cost for finite length polymers. The assumption of a harmonic tube was confirmed by simulation results. By decomposition into independent stiff rods of appropriate length, we were able to establish an implicit equation for the deflection length. The resulting tube width $L_\perp$ is in agreement with the established scaling law $L_\perp = c\xi^{6/5} l_\mathrm{p}^{-1/5}$ with mesh size $\xi = \sqrt{3/\nu L}$ and persistence length l_p. Our theory provides a prefactor of $c \approx 0.31$ and a higher order term that accounts for finite length effects and scales with ξ^2/L. These finite length corrections additionally allow to quantitatively account for polydisperse networks.

The available experimental data is consistent with our predictions. Since there are additional intrinsic collective fluctuations in experiments, the theoretical estimate provides a lower bound. To provide a precise validation, we have complemented our theoretical work by extensive Monte Carlo simulations of a test polymer in an environment of obstacles. The resulting self-consistent tube widths perfectly match the theoretical value predicted. This strongly supports the validity of the absolute value for the concentration dependent tube diameter.

Furthermore, we have employed simulations to observe properties beyond the analytical theory. We have recorded the distribution function of tube widths in a network for different concentrations. Thereby we were able to explicitly confirm self-consistency of the simplifying model with a fixed tube diameter.

Both our theoretical predictions, e.g. the finite length contributions to the tube diameter, and our simulation data, e.g. the distribution functions, provide the opportunity of feasible experimental comparison. On the theoretical side, the significance of correlations and collective fluctuations of the complete medium, as well as an analytical model of distribution functions may open up promising continuations of this work.

4.1 Tube Model

We consider entangled solutions of semiflexible polymers where chemical bonds by cross-linking proteins are ruled out. While F-actin is a prominent example of this class of biopolymers and has a strong record of experimental data available, the work is furthermore also applicable to other semi-flexible polymer solutions in a comparable regime where a description in the realm of the tube model is justified [83]. As binding by cross-linking proteins is ruled out, the only inter-polymer interactions are the topological constraints that are imposed by the fact that the network constituents cannot cross through each other. The polymers are considered to be mathematical lines as their thickness is negligible [92] and no noteworthy long-ranged attractive or repulsive interactions exists. Consequently the system has no excluded volume. Typical F-action solutions are polydisperse with a mean filament length $L \approx 22\mu\text{m}$ [84]. The detailed length distribution however is highly variable for different preparations [48, 89] and we will consider monodispersity in the following. With a persistence length $l_{\text{p}} \approx 17\mu\text{m}$ [22, 85] comparable to its length F-actin is a typical semi-flexible polymer. The polymer's bending stiffness κ is related to the persistence length as $l_{\text{p}} = \kappa/k_{\text{B}}T$ and each polymer's configuration $\text{r}(s)$ is parameterized by the arc length s. A free polymer then is described in the worm-like chain model [24, 25] by the Hamiltonian

$$H(\kappa) = \frac{\kappa}{2} \int_0^L ds \left(\frac{\partial^2 \text{r}(s)}{\partial s^2} \right)^2 , \tag{4.1}$$

where the second derivative of $\text{r}(s)$ is the local curvature $\mathcal{C}$ at arc-length s. Consequently, the distribution of local curvatures of a free polymer is

$$P(\mathcal{C}) \propto \exp\left(-\frac{1}{k_{\text{B}}T} H(\kappa, \mathcal{C}) \right) \propto \exp\left(-l_{\text{p}} \mathcal{C}^2 \right) \tag{4.2}$$

and the resulting Gaussian distribution's width decreases with increasing persistence length of the polymers. The density ν of a network of these polymers is given by the number of polymers of length L per unit volume. At a concentration of $c = 0.5$ mg/ml corresponding to $\nu \approx 1\mu m^{-3}$ [86] the average distance to the next neighbor is given by the mesh mesh size $\xi \approx (\nu L)^{-1/2} \approx 0.2\mu\text{m}$ and therefore much smaller then polymer length and persistence length, $\xi \ll l_{\text{p}}$. Due to this ratio of length scales it is guaranteed, that a specific polymer will not deviate far from its average contour and it is highly unlikely to fold back onto itself. Thus it is feasible to model the combined effect of all neighboring filaments of an arbitrary probe polymer by a hypothetical tube potential. The tube potential has an harmonic profile as observed in experiments and simulations [90, 93]. Due to the disorder in the network the tube diameter and thus the local strength of the tube potential vary along the contour [57]. The tube is conventionally described by a potential strength $\gamma(s)$ that is parameterized by the arc length s along the tube backbone or tube contour given by the potential's minimum in space. If we denote this tube backbone by $\text{r}_0(s)$, the resulting energy becomes the sum of the bending energy of the polymer and its confinement by a harmonic potential around the tube backbone

$$H(\gamma, \kappa) = \int_0^L ds \left[\frac{\kappa}{2} \left(\frac{\partial^2 \mathrm{r}(s)}{\partial s^2} \right)^2 + \frac{\gamma}{2} \left(\mathrm{r}(s) - \mathrm{r}_0(s) \right)^2 \right] . \tag{4.3}$$

In contrast to a free polymer this equation does not allow to infer a simple distribution of curvatures as in Eq. 4.2 because the second term causes an additional dependence on the tube contour. This term causes a confinement around the minimum of the potential that is given by the tube backbone as explained above. If for instance the backbone is already strongly bend and the confinement potential is sufficiently strong, high curvatures are more likely than for a free polymer. Obviously the distribution of curvatures of any probe polymer sensitively depends on the actual form of the tube to which it is confined. Furthermore, this tube backbone is itself a statistically distributed quantity that only results from the initial configuration of the network and is not a priori know. Finally, it is important to state that this tube is only well-defined up to intermediate time scales. Therefore in networks of non cross-linked polymers the tube model cannot be used as an equilibrium concept without further thought. The confinement tube is defined as the space accessible to an encaged polymer in an environment of neighboring filaments before large scale reconstruction of the network changes this environment. The tube picture is thus a valid description as soon as the polymer experiences topological interaction with its neighbors and as long as these are not remodelled by large scale dynamics. The first time scale can be estimated from dynamic light scattering as the point were a cross-over from free filament to restricted dynamics sets in at about 10 ms [65]. A measure for the time scale of remodelling can be obtained from the time it takes the probe filament to leave its initial tube. For unstabilized actin filaments this process is dominated by treadmilling occurring at an approximate rate of 2 μm per hour [94]. Stabilized actin filaments where treadmilling is abolished by phalloidin can only reptate out of their tubes at much slower rates. Reptation rates in this case have been estimated from experiments [66, 67] to be as long as several days for a 10 μm long filament.

As explained above, a well established scaling law for the average tube diameter has been derived by Semenov. Recently, also theories providing absolute values have been proposed [93, 80] and confirmed by experiments [57]. In analyzing experimental data it is important to keep in mind that e.g. fluorescent microscopy only provides an observation of an effectively two dimensional focal plane. For the tube diameter for instance, this signifies that only the fluctuations in one Cartesian component are measured. In the focal plane the system can be considered as a probe filament surrounded by fluctuating point obstacles that represent the cuts of neighboring polymers through the the plane of observation as depicted in Figure 4.1.

4.2 Monte-Carlo Simulations

Motivated by this point of view, we have developed a Monte-Carlo simulation with the standard Hastings-Metropolis algorithm [95] to observe curvature distribution functions

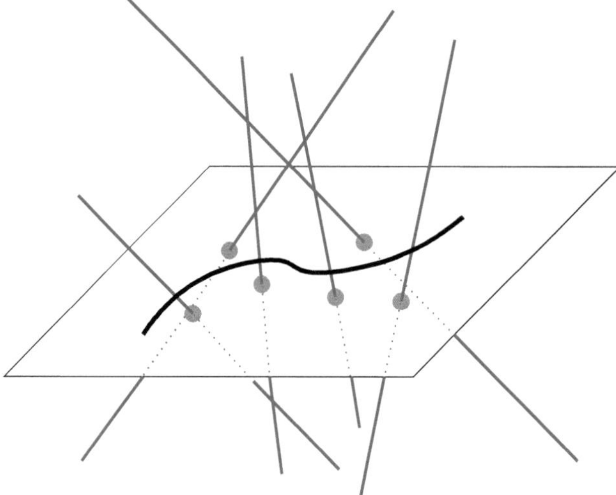

Figure 4.1: In a two-dimensional cut through the network, as e.g. in the focal plane of a microscope, neighboring polymers reduce to point-like obstacles.

of a single probe filament in a two-dimensional plane of point-like obstacles. This does not only provide ready comparability with experiments, but is also a valid simplification of the three-dimensional problem since transverse undulations in different components can be assumed to be independent. As the surrounding network constituents are very thin and nearly straight due to their large persistence length it is justified to represent them by point-like obstacles. Those obstacles undergo themselves transversal fluctuations that have to exhibit on average the same characteristics as the fluctuations of the probe filament for reasons of self-consistency. As the latter were found to be harmonic, the same must hold for the in-plane fluctuations of an obstacle polymer that cuts perpendicularly through the simulation plane. If the cutting angle is tilted the harmonic profile is distorted corresponding to different fluctuation strength for the two in-plane components. While the average tube diameter must equal the fluctuation strength averaged over all obstacles, the distribution of these two quantities may be broad [93]. It was found however, that the simulation results are rather insensitive to these parameters. The simulation is thus an adequate and self-consistent description of the physical problem of a entangled network of semi-flexible polymers, if the parameters obstacle density ρ and obstacle fluctuation potential strength γ are chosen to represent the corresponding polymer density ν and resulting tube diameter $L_\perp$. These parameters have been shown [93] to be:

$$\rho = \frac{2}{\pi}\nu L \,, \qquad \gamma \approx 2.78(\nu L)^{6/5} l_p^{2/5} \,. \tag{4.4}$$

The probe filament of length L is initially placed in a straight configuration onto the plane of observation and is represented by a sequence of N connected segments with orientation t_i. Due to inextensibility the segments are of fixed length L/N. In a first step the filament is allowed to relax on the plane without the presence of any obstacles. The relaxation is performed with respect to an Hamiltonian :

$$H(\{\mathrm{t}_i\}) = -J\sum_{i=1}^{N-1} \mathrm{t}_i \mathrm{t}_{i+1} \,, \tag{4.5}$$

where in two dimensions the relation between persistence length and J is given as $l_p/L = -(N\ln(I_1(J)/I_0(J)))^{-1}$ with I_n the modified Bessel functions of first kind [96]. After equilibration with respect to the Hamiltonian (4.5) the probe filament features the bending distribution of a free polymer. Now the obstacle fluctuation centers p_j^0 are fixed to random positions of the simulation plane. While these centers remain fixed for the course of the simulation, the point obstacles p_j themselves - initially placed at $p_j = p_j^0$ - are allowed to move in a harmonic potential $U = \frac{1}{2}\gamma_j(p_j - p_j^0)^2$. Their motion is not only governed by this potential but also by the constraint that they must not cross the probe filament and they remain on that side of the probe filament where they had been initially placed [1]. Naturally,

[1]We also performed simulations where the topology of the obstacles, i.e. the side of the filament they are constrained to, is only determined after they are allowed to relax away from their fluctuation center. Thereby initial conditions with the probe filament lying between an obstacle and its fluctuation center become possible. The presence or absence of these "misfit"-configurations does not change our results.

the same constraint also holds for the probe filament where every move, that would lead to a configuration where an obstacle point had switched sides, is rejected. While the motion of the point obstacles is straightforward, we will discuss the moves of the probe filament in more detail in the following section.

4.2.1 Dynamic Trial Moves

In the construction of trial moves our intention has been to find a set of moves that mimics the underlying dynamics in the physical system as closely as possible. To this end it is of particular importance to keep the relevant time scales in mind. Therefore, our choice of moves describes the effect of the underlying physical forces and dynamics on a probe filament for times well below large scale rearrangement of the network. These have to include transverse undulations, exploration of void spaces along the contour and small-scale reptation and breathing, while the effects of large scale reptation like annihilation and creation of obstacle points remain impossible. In the following, we will thoroughly explain the moves used in our Monte-Carlo simulations. As shown below, the resulting data for conformations and distribution functions sensitively depends on their nature.

We use four different moves, that change the polymer configuration on different levels ranging from local change in only one tangent up to a global manipulation of the complete polymer contour. The classical random pivot move (see Figure 4.2 (a)) chooses a random tangent AB and rotates it by a small random angle. This changes the two bending angles at A and B but leaves all other angles invariant. However, all positions along the polymer contour are modified - predominantly in a direction transversal to the contour. Another move employed is a flip move (see Figure 4.2 (b)). Here two beads A and B separated by a random number of segments are picked and all points are mirrored along the axis connecting A and B. This can be seen as the two-dimensional analogon to a crankshaft move. The flip move leaves the ends of the polymer unaffected. These two rather common moves are known to effectively explore the available phase space of free polymers or even of polymers in pore-like potentials. However, they lack the ability to mimic the motion of a polymer into the void spaces between obstacles along the walls of the hypothetical tube.

To this end we use novel moves that are depicted in Figure 4.2 (c) and (d). They simulate the exploration of a local void space, i.e. a part where the "tube" formed by the fluctuating obstacles is rather large. Due to the negligible longitudinal extensibility of semi-flexible polymers the motion of the polymer into this void space is obviously possible only if the other parts of the polymer are retracted or straightened out. Let us first introduce the move that performs the latter and is illustrated in 4.2 (c). The additional length that is needed to enable the protrusion into a void space is obtained from undulations in adjacent parts of the polymer. A stronger bending in one part is made possible by weaker bending in other parts and we therefore choose the label "trade-off" move. Specifically, the move is conducted by randomly choosing two points C and E, that are separated by an even distance of segments (in our simulations we apply two different moves where this distance is either two or four). This is the region of the polymer that should explore the void space. Furthermore, to the left and right of this region two more points are chosen that enclose

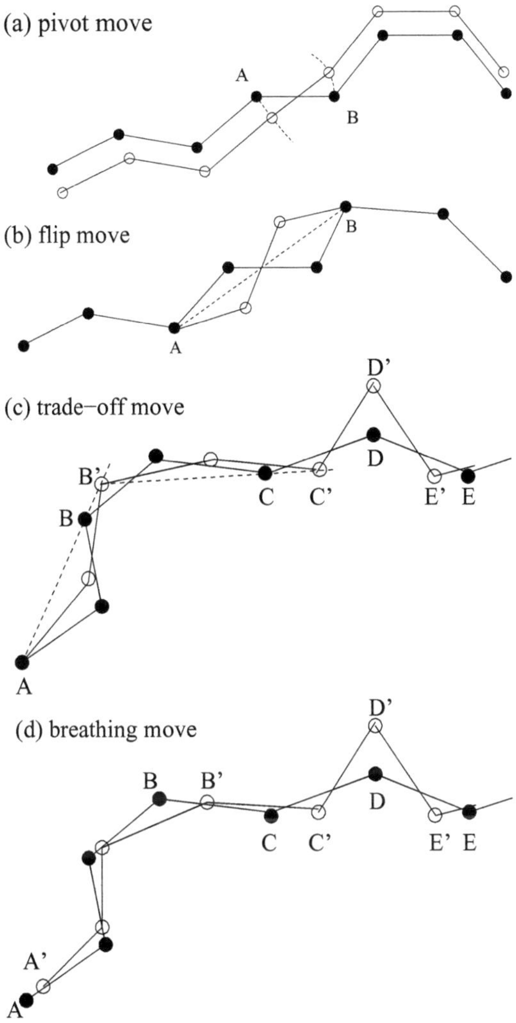

(a) pivot move

A B

(b) flip move

B A

(c) trade–off move

D' D B' B C C' E' E A

(d) breathing move

D' B B' D C C' E' E A' A

Figure 4.2: The different moves performed during the simulations. Full circles denote the original positions and open circles the new configuration: (a) The pivot move changes two angles and all positions. (b) The flip move changes all angles and positions between A and B. (c) The "trade-off" move changes the curvature of the section CE in a trade-off with the adjacent section to both sides (only l.h.s section AC depicted here). (d) The "breathing" move changes all positions and angles and axially moves the polymer's ends.

the regions that will be straightened out to compensate for the contour length drawn into the void. At the left hand site this is for example point A that is separated from C by an again even number of tangents (in our simulations two to ten). We restrict our illustration to the left hand site, but the same process also applies to a comparable region on the right hand side that is not shown. In addition to a random number R_{loc} to choose the location of the section CE along the contour, we draw another random number R_{δ} that quantifies the extend of change in bending. Our goal is now to construct a algorithm that straightens out the region AC in order to enable the region CE to realize a stronger bending. In a first step the vector AB is prolonged to AB' by adding a random small amount δ. The new point between A and B is chosen in a way to keep tangent length conserved. In a next step B'C is prolonged to B'C' by adding the same δ and so on. The same process is performed on the right hand side tail. This results in two points C' and E' that are separated by a smaller distance than the original pair C and E. The remaining section in between is now fitted in under the precondition of length conservation. After construction the new configuration is checked for violation of the topological constraints and bending energy. Naturally, also a backward move has to be possible that pushes back the region CE into a lesser bend conformation by creating or enhancing undulations in the region AC. This process is performed if the random distance δ is negative and is achieved in the following way. We proceed in a reverse fashion by first choosing points B' and C' and reducing their distance by δ. The same is performed along the tails until the final point of the region AC is reached. The result is a distance between points C and E that is now larger than the original distance between the points C' and E' and thus a section CE that is pulled back from the void to a straighter configuration and enhanced undulations along the left and right tails. Note that exclusive performance of this move will leave the polymer ends unchanged. In combination with the other moves however, it allows for a longitudinal motion of the ends by effectively manipulating the undulations along the encaged contour. After having constructed a novel trial move for a Monte-Carlo simulation it is of crucial importance to check if it guarantees a conversion of the simulation to equilibrium. While it is know that this can be achieved by the balance condition [97] it is usually more convenient to check the stricter condition of detailed balance [31]. It requires that in equilibrium for every pair of configurations m and n the moves from m to n given by $P^0_m P^{\mathrm{move}}_{m\to n} P^{\mathrm{acc}}_{m\to n}$ equal the number of reverse moves $P^0_n P^{\mathrm{move}}_{n\to m} P^{\mathrm{acc}}_{n\to m}$. Here, P^0 is the Boltzmann weight of a configuration, P^{move} is the a priori probability to select a certain move and P^{acc} is the probability that this move is accepted. For the move introduced move above, every move is characterized by the two random numbers R_{loc} and R_{δ} and its backmove is simply obtained by changing the sign of δ. Consequently, the a priori probabilities of every move and its backmove cancel. The acceptance probability is zero for both move and backmove if topological constraints are violated. If the topology is conserved the move is accepted according to change in bending energy and thus according to the ratio of Boltzmann factors and consequently the condition of detailed balance is guaranteed.

Finally the fourth trial move, a global "breathing" move, mimics a global retraction of the polymer along its contour to enable the exploration of void spaces. This move causes a pronounced axial motion of the polymer ends and is schematically depicted in Figure

4.2 (d). Again, by a random number $R_{\rm loc}$ a section CE is randomly chosen in which the local curvature is to be manipulated. Either the bending of CE is enhanced resulting in a global retraction of the remaining tail sections of the polymer to the left and right hand side of CE, or the bending of CE is diminished which is achieved by pushing out the remaining sections. As the manipulation of the tail sections is always to occur along the polymer's contour, bending of CE causes axial motion of both ends towards the polymer's center and straightening out of CE causes end motion away from the center. In detail the new configuration in the former case is constructed by choosing a new C' by reducing the distance between C and E by a small random δ as above. B' is then found at that point where a radius L/N intersects with the old polymer contour. All other points are chosen accordingly proceeding towards the polymer's ends. The back move is obtained by extending the section CE by δ and fixing the direction of the tail segments by demanding that they pass through the joint points of the old contour [2]. Thereby, again the exact backmove to any given move is simply obtained by inverting the sign of δ. Consequently, the same reasoning as above also proves detailed balance.

We validated this particular choice of Monte-Carlo moves in a simulation of a single free polymer, where we compared our observations to established results of the bending distribution of free polymers, tangent-tangent correlation function and end-to-end distribution function. All results presented in the following were obtained as a combined ensemble and time average. Ensemble averaging was performed over a large number of initial obstacle fluctuation center distributions. Additionally, for every initial obstacle distribution several initial distributions for the probe filament were chosen. This can also be seen as averaging over different topologies. After initial equilibration, observables were monitored and averaged for the remainder of the simulation time thereby averaging over all statistically allowed configurations in a fixed topology.

4.2.2 Simulation Results

In a first step we characterized the conformation of the confinement tubes - a quantity that is also accessible by fluorescence microscopy and therefore allows for a comparison to experimental data. To this end we determine the tube contour, i.e. the backbone of the area to which the probe filament is confined, by averaging over the contour of the probe polymer in its cage of point obstacles over the evolution. From the resulting contour we determine a curvature distribution $P(\mathcal{C})$, where the curvature is defined by locally fitting a parabola with $y = \mathcal{C}/2x^2$ to the contour. The distribution obtained after averaging over initial conditions is shown in Figure 4.3.

We compare our data to measurements that were obtained by fluorescence microscopy [57] of in vitro solutions of rhodamine-phalloidin labeled F-actin on a minute time scale [3]. The good agreement of the data is evidence that our simulation approach provides a

[2]E.g. in going back from C' to C, the new point B is found by choosing the direction of the new tangent BC to pass through B'

[3]The experimental data was obtained from measurements at several concentrations. While the observed effect is in principle dependent on concentration as we will discuss below, simulations show no significant

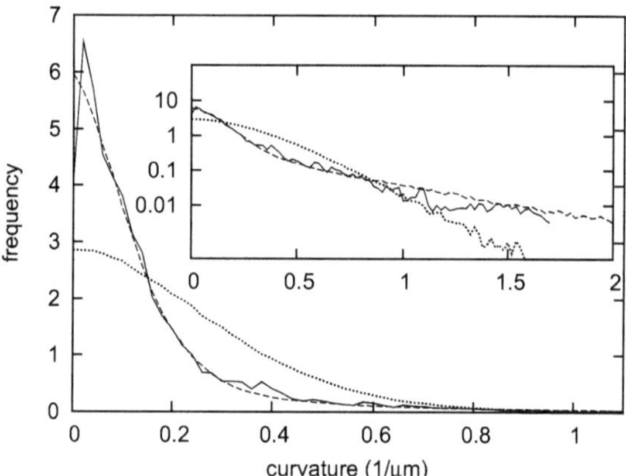

Figure 4.3: Curvature distribution of confinement tube contours as obtained from Monte-Carlo simulations (dashed line) and from experiments (full line) [57]. The curvature distribution of a free filament obtained by the same simulation algorithm is plotted for comparison (dotted line). The inset shows the same data in a semilogarithmic plot.

reliable representation of the physical system under consideration.

In comparison to the curvature distribution obtained by the same algorithm for free filaments, two distinctive differences emerge. While the free filament distribution has to be Gaussian as explained in Eq. 4.2, the distribution function of the tube contours features a pronounced exponential tail. As an exponential decays much slower as a Gaussian towards high values, this signifies that highly bend filaments are much more frequent. Also for the occurrence of small curvatures a dramatic increase in probability compared to the case of free filaments can be observed. The form of the distribution however remains Gaussian, making the difference rather quantitative. It is obvious that relative to free filaments, probability is both shifted to smaller and larger curvatures at the cost of medium curvatures. This reflects the visual observation that tube contours are on average straighter than free polymers but also feature distinctive strongly bend sections. The first feature, i.e. the increase of small local bendings, obviously results only from the averaging procedure carried out in determining the tube backbone. The averaging over all topologically allowed polymer conformations within the tube integrates out fluctuations of small wavelength (small radii of curvature) to obtain a larger radius of curvature for the coarse-grained tube contour. The increase in highly bend filaments on the exponential tail of the curvature distribution is far less obvious. To avoid the complications of coarse-graining related to the tube contour, we turn to a different observable. The curvature distribution of the encaged filament itself is not as easily accessible to experiments and thus does not allow for verification, but it allows a direct comparison to free filaments. In particular, this is the case for solutions with negligible excluded volume, where the bending distribution obtained by standard statistical mechanics should be identical.

We recorded snapshots of the probe polymer during the evolution and analyzed these for their curvature as explained above to obtain the curvature distribution of confined filaments depicted in Figure 4.4. Now the distribution at small and medium curvatures remains largely unaffected as compared to free filaments. However, the pronounced exponential tail at high curvatures is still observed. These features were observed for networks composed of polymers of different persistence length. At a first glance this behavior seems to contradict general concepts of statistical mechanics that do not predict any effect of topological constraints in a system without excluded volume. We will show in the following how this conflict is resolved.

4.3 Thermodynamic Interpretation

As discussed in Sec. 4.1, the probe filament is confined to its tube during the time window that is relevant to many biological processes and that is the observation frame for most experimental measurements as well. Clearly, our simulations have also been tailored to represent this intermediate time scale as point-obstacle centers are fixed and large scale reptation is beyond simulation time. Therefore all observations made on this time scale are

dependence in the experimental range of actin concentration. We thus chose to combine the data of different concentrations for the sake of a smaller statistical error.

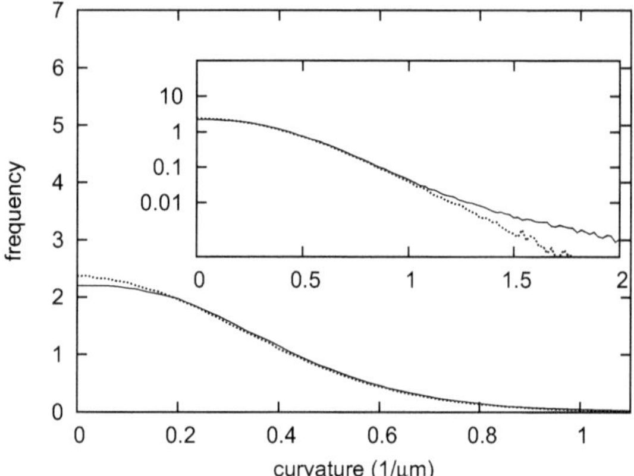

Figure 4.4: Curvature distribution of encaged filaments obtained from Monte-Carlo simulations (solid line) compared to the distribution of a free filament (dotted line). The inset shows the same data in a semilogarithmic plot.

crucially influenced by the tube's properties and we will discuss the implications for the obtained averages and trace back the results for the deviation of the curvature distribution from the free filament case. To this end, we will first of all recapitulate some general notions on standard thermodynamic averaging, then pinpoint differences to averaging procedures in the tube model and finally present additional simulations to corroborate our findings. To facilitate our discussion we restrict ourselves in the following to the case of a probe filament in a two-dimensional array of fixed point-like obstacles. This simplified system has the same general characteristic as a polymer confined to a tube in a network but considerably less degrees of freedom.

4.3.1 Ensemble Average - Time Average

If one is faced with the problem to calculate averages for a statistical system there are in general two different possibilities: an ensemble average and a time average that will yield the same result if the system is ergodic. An ensemble average in the system under consideration could be realized by drawing a large number of allowed configuration of the complete network and weighing them by the corresponding Boltzmann factor. As there is no interaction between polymer and obstacles and in the absence of excluded volume no initial configuration can be rejected due to hard-core exclusion, the only contribution to the Boltzmann factor is the bending term of the worm-like chain (Eq. 4.1) and the average obtained has to equal the case of free polymers. Due to the absence of excluded volume the polymer is not able to see the obstacles and any probe polymer inserted into the network has zero chance of overlap or rejection. The concept of a time average on the contrary would be to start from one initial probe polymer configuration and monitor the following time evolution. As soon as the probe polymer has explored every point in phase space, the obtained average equals the ensemble average. In the obstacle system however, the time to fulfill this requirement is exceedingly long. Points that might be very close to each other in phase space can be very far apart in terms of transition time. This is due to the fact that the topological constraints that the obstacles impose, partition the phase space into a multitude of areas that are not directly connected. Consequently, the probe filament can only traverse a point obstacle by completely reptating back and forth. Due to the immense number of different topologies and the slow reptation time scale (see Sec. 4.1) a complete time average is not only out of the scope of simulations and experiments but also irrelevant to biological processes.

4.3.2 Partitioned Averaging

It can now be tried to substitute the infeasible sampling of phase space by means of reptation of a single test polymer by a large number of samples with different initial conditions. Here, initial configurations of the test polymer are drawn from the free polymer distribution and randomly placed into the obstacle network. Different topologies emerge as the same obstacle could be at the left or right of the test polymer. Starting from these initial conditions the polymer's configurations are now sampled employing a Markovian

Monte-Carlo dynamics respecting the topological constraints imposed by the neighboring obstacles. Such a procedure corresponds to a partitioning of phase space into sections with "different topologies", i.e. areas that are not directly connected. One could therefore argue, that an average containing all possible topologies should also hold the same results as a complete time average or averaging of a free polymer. However, this argument can only be valid if the partitions of phase space are disjunct. Otherwise if these partitions overlap, the averaging procedure will put a higher or smaller weight on some microstates. The curvature distribution obtained by the experimental and simulational averaging procedure can thus only be expected to equal the free polymer case, if it is guaranteed that during the observation time the topology and hence the partition of phase space remains unaltered for every test polymer. Processes that can modify the topology are for instance reptation or "breathing" of the polymer. One mutual feature of these processes is that they cause motion of the polymer ends tangentially along the contour. Therefore obstacles can switch their topology with respect to the test polymer, e.g. an obstacle initially left of the test polymer can end up being on the right side after the polymer's end has moved back and forth. The requirement of strict disjunct partition of phase space would thus essentially amount to the constraint of keeping the polymer's ends fixed which is evidently not the case in the actual physical system.

We therefore conclude that the polymer dynamics inside the confinement tubes are metrically transitive due to their characteristic features as breathing and reptation that change the topology in the network array. The topological partitioning is thus not maintained under the dynamic evolution. Therefore the resulting averages and distribution functions have not necessarily to equal the corresponding results for free polymers. This holds on intermediate time scales before large scale reptation sets in, which are the time scales of experimental observation. In the long time limit however, when the single polymers of the ensemble have been able to explore larger parts of the phase space beyond their confinement tubes, free filament distributions should be recovered. Consequently, the observed non-equilibrium distribution functions do not violate thermodynamic requirements as they are transient. However, the time needed for total equilibration is so long, that it is not reached on any applicable time scale.

After we have shown, how transient non-equilibrium distribution functions can arise on intermediate time scales even in the absence of excluded volume, we will use additional simulations to clarify the physical origins of highly bend filaments.

4.3.3 Additional Simulations - Entropic Trapping

To this end we have conducted further simulations of the simplified system of fixed obstacles. Note, that this work does not apply to a network of F-Actin as represented by the simulations with self-consistently fluctuating obstacles. It merely serves as an model system for our considerations on the averaging procedures. First of all, we have checked if a system where dynamics have chosen to be metrically intransitive faithfully reproduces the distribution functions of free filaments. This was achieved by running simulations where the filament's Monte-Carlo moves are restricted to "flip" and "trade-off" moves. This

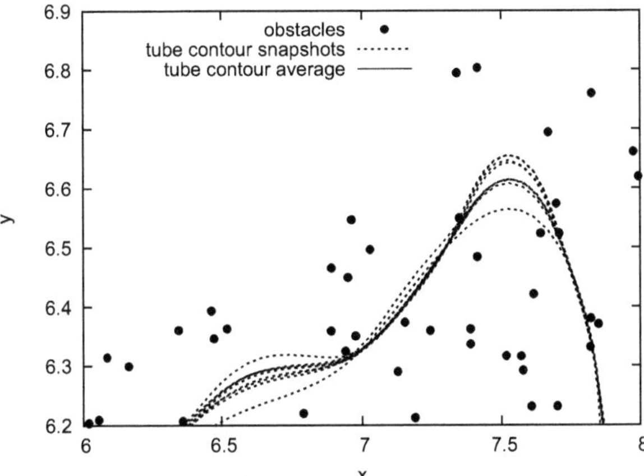

Figure 4.5: A typical network configuration where transient entropic trapping occurs when the probe filament explores a void space by high bending thereby realizing an entropic gain.

choice ensures that the polymer's ends do not move axially. As obstacles remain fixed, it is guaranteed that the topological partitioning cannot change. The resulting distribution function indeed reproduces the case of free filaments (see Figure 4.6 (top)). Furthermore, we have identified the physical origins of the highly bend parts of the test polymer. It turns out that high curvatures occur always at local initial topologies where a test polymer can protrude into a large void space in the obstacle array (see Figure 4.5). The initial conformation of the test polymer already has a curvature that facilitates a further bending into a large void part in the network. Hereby the system realizes a higher entropy by bending harder than the equilibrium curvature distribution. These events are rare but they dominate the tail of the curvature distribution. Apparently, the polymer is trapped in these entropically favorable configurations on the time scale of observation. This behavior bears some similarity to "entropic trapping" observed for flexible polymers in random environments [98, 99, 100, 101]. Note in particular, that these conformations also result in a pulling back of the polymer ends and thus a change of phase space partition. Hence, this observation does not only clarify the physical cause of high bendings but also proves according to the argumentation above that a curvature distribution different from the free polymer distribution does not violate statistical mechanics. Furthermore, we have investigated how these special conformations are realized as a function of the fluctuation amplitude of the obstacles. For both self-consistently fluctuating and immobile obstacles an exponential tail is visible in the curvature distribution. This is also the case for obstacles fluctuating with a higher amplitude as in the self-consistent case. The weight on the high curvature tail is highest for immobile obstacles and decreases with increasing fluctuation amplitude. This is consistent with the explanation for the high bendings provided above. As the effective size of void spaces is diminished with larger obstacle fluctuation the effect decreases. In the limiting case of very large fluctuations network obstacles become delocalized, the system is reduced to a gas and the curvature distribution of a free polymer will be recovered.

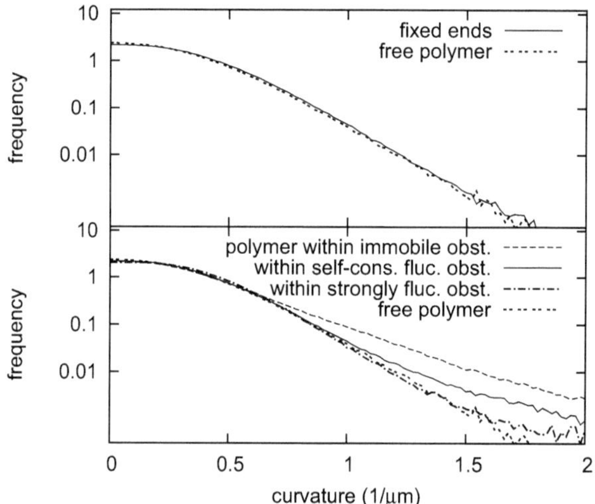

Figure 4.6: Curvature distributions compared to free filaments (dotted line): (top) metrically intransitive system with fixed ends (solid line), (bottom) probe filament in obstacles of different fluctuation strength.

4.4 Conclusion

We have investigated the curvature distribution functions in entangled networks of semi-flexible polymers. To this end we developed an approach to simulate single probe filaments in an entangled network of semi-flexible polymers by a self-consistent reduction to a two-dimensional setup corresponding e.g. to the focal plane of a microscope to allow for comparison for experimental data. This Monte-Carlo simulations were particularly designed to mimic the real polymer dynamics on intermediate time scales by allowing for an effective exploration of network void spaces by breathing and short scale reptation. The simulations provide data on curvature distributions for tube contours that agree well with fluorescence microscope measurements on F-actin solutions [57]. Furthermore, they permit to observe curvature distributions of single confined filaments. These distributions feature an unexpectedly high weight on highly bend filaments that is traced back to transient entropic trapping in network void spaces. The fact that the equilibrium distribution of free polymers is not recovered even in the absence of excluded volume, is shown to be an immanent feature of the polymer dynamics in a disordered environment on intermediate time scales below large scale reptation. The fact that this regime is best described by the tube model - a non-equilibrium concept - explains that a treatment in terms of equilibrium thermodynamics is inappropriate. Consequently, the observation of transient non-equilibrium distribution functions is a generic effect observed for all measurements on time scales both relevant to experiments and feasible for simulations. These findings provide insight into the conformation of confined polymers and can e.g. prove useful for further analysis of reptation or emerging collective macroscopic properties.

Chapter 5

Non-Affine Deformations

5.1 Introduction

After we have analyzed the equilibrium tube diameter in Chapter 3 and pointed out the limitations of the tube model as an equilibrium concept in Chapter 4, we now want to focus on non-equilibrium properties of entangled networks of semiflexible polymers. Here, our central goal is to find an appropriate description for the processes inside a network that occur if the system is brought out of equilibrium as for example by a shear deformation. The mechanical response of the network to oscillatory shear can be observed experimentally and shows a rich dependence on frequency [74, 75, 80, 53]. Furthermore, upon application of larger stress a non-linear regime was investigated and shear stiffening of the network was observed [102, 103]. While details of the microscopic deformation field remain unclear, different coarse-grained theoretical descriptions have tried to grasp the emergent properties. Regarding the scaling of the plateau modulus G with concentration c in the regime of linear response Isambert and Maggs [79] derived a power-law of $G \propto c^{7/5}$ from a simple scaling argument based on the deformation of confinement tubes as explained in Chapter 2. Other theories attribute the material response to the suppression of undulations by stretching [32, 29]. They predict a scaling with a considerably larger exponent, but seem to disagree with experimental data [50, 78, 104]. Recent experimental work [105] claims to have verified an exponent of 4/3 predicted by an elastic medium theory [80]. Common to all these approaches is the assumption that the macroscopic deformation field is assumed to be affinely transmitted to all length scales. It is however rather unlikely that this strong simplification correctly describes the network response at a microscopic level. Recently, the importance of non-affine deformations in cross-linked networks was pointed out [106]. In this chapter we will present an approach that permits to go beyond the assumption of affine displacement and investigate resulting differences. Additional our numerical results permit for the first time to distinguish between the scaling laws predicted by competing theoretical models.

We proceed as follows: in Section 5.2 we define the system under consideration. We introduce the Hamiltonian of the network, review the simplifications that lead to the tube model and prior work on the plateau modulus, and emphasize the mean-field nature of the tube model and the assumption of affine displacement. In Section 5.3 we present our approach to numerically compute the free energy of the system by a reduction to two dimensions. We explain the effect of global shear on the microscopic constituents of the network and introduce a free energy minimization procedure that results in a non-affine deformation field. We proceed in Section 5.4 with the presentation and interpretation of our results before we conclude in Section 5.5.

5.2 System Definition

We consider a network of semiflexible polymers that only interact via a hard-core potential and thus constrain each other topologicaly. The polymer density is given by the number ν of polymers of length L per unit volume. The stiffness κ of the polymers gives rise to a persistence length of $l_p = \kappa/k_\mathrm{B}T$. The configuration of the i-th polymer in space, $\mathrm{r}_i(s)$, is parameterized by arc length s and the average distance between the networks constituents is characterized by the mesh size, $\xi = \sqrt{3/\nu L}$. Concerning the mechanical response of the network, we are interested in the dynamic processes that occur after a deformation has driven the system out of equilibrium. These relaxation processes occur on different time scales. While generally every stress can relax by reptation of the polymers, this process is dramatically slowed down due to topological constraints in crowded environments [107]. On intermediate time scales relevant for the plateau modulus it can be assumed that the constraints imposed by surrounding polymers can not be overcome and that the center of mass of all filaments does not change substantially. A given polymer i is then described by the worm-like chain model [24, 25] with a Hamiltonian that has contributions from intra-polymer bending and from interactions with the neighboring filaments in the solution. This can be formally written as

$$H_i = \frac{\kappa}{2}\int_0^L ds \left(\frac{\partial^2 \mathrm{r}_i(s)}{\partial s^2}\right)^2 + \sum_{j\neq i} \Theta[\mathrm{r}_i(s), \mathrm{r}_j(s)] \,, \tag{5.1}$$

where the function $\Theta[\mathrm{r}_i(s), \mathrm{r}_j(s)]$ describes the hard-core interactions between the polymers i and j. Note that this function also depends on the initial topology of the network and is only zero as long as the two polymers do not cross each other. If the initial topology is changed by an interpenetration of two polymers, $\Theta[\mathrm{r}_i(s), \mathrm{r}_j(s)]$ would return an infinite energy. The Hamiltonian of the complete system is obtained as $H = \sum_i H_i$.

Given the form of the Hamiltonian (5.1) a calculation of the free energy F from the partition sum Z as $F = -k_\mathrm{B}T \ln Z$ is obviously not feasible as the partition sum

$$Z = \Pi_i \int \mathcal{D}[\mathrm{r}_i(s)] \exp[-H/(k_\mathrm{B}T)] \tag{5.2}$$

amounts to multiple path integrals of a highly irregular integrand.

A simplifying description of the system was proposed with the famous tube model [54, 55] where the combined effect of the fluctuating neighbor polymers on a single test polymer is described by an effective harmonic potential. While the tube model is the foundation for theories for different properties of semiflexible polymer networks like tube diameter [60, 80, 93] or viscoelasticity [79], it has to be kept in mind that it only provides a mean-field description of the microscopic constituents. In the remainder of this section we will review how free energies and mechanical properties can be derived from the tube model and point out possible shortcomings of this coarse-grained frame of description.

The tube model can be applied to solutions of semiflexible polymers were the confinement of a single polymer by its neighbors is sufficiently strong to guarantee that the transversal undulations of the polymer do not deviate far from an average contour in space - the tube backbone $\mathrm{r}^0(s)$. This is the case, if persistence length and polymer contour length are substantially larger than the typical void spaces in the mesh of surrounding polymers, thus $L, l_p \gg \xi$ as e.g. given for most F-actin networks. The complex sum in the second term of the Hamiltonian (5.1) can then conveniently be substituted by a harmonic potential with average strength γ and minimum at the tube backbone:

$$H(\gamma, \kappa) = \int_0^L ds \left[\frac{\kappa}{2} \left(\frac{\partial^2 \mathrm{r}(s)}{\partial s^2} \right)^2 + \frac{\gamma}{2} \left(\mathrm{r}(s) - \mathrm{r}_0(s) \right)^2 \right] . \tag{5.3}$$

As pointed out by Odijk [58] it is instructive to introduce an additional length scale $L_d \approx (\nu L)^{-2/5} l_p^{1/5}$ known as deflection or Odijk length. While the length scales L, l_p describe the properties of the single polymer and the length scale ξ describes the network, the deflection length L_d captures the interaction between both. It can be interpreted as a measure for the distance between two collisions between the encaged polymer and the tube walls and therefore the number of collisions of a polymer is given as L/L_d. This is also reflected in the free energy cost ΔF that arises from the restriction of the test polymer to a tube and is obtained by a path integration of (5.3) over all polymer configurations [59] as

$$\Delta F = \sqrt{2} k_{\mathrm{B}} T \frac{L}{L_d} . \tag{5.4}$$

This signifies that at a spacing of L_d between two collisions every of the L/L_d contact points between polymer and tube contributes one $k_{\mathrm{B}}T$ to the confinement free energy.

Having derived the free energy in the coarse-grained tube model, the next step is to analyze the change in free energy at mechanical deformation to obtain the plateau modulus. As the polymers are described in terms of their tubes, it is obvious to investigate the effect of deformation on the tubes for which the free energy is known as reasoned above. Together with a scaling law $d \propto c^{-3/5} l_p^{-1/5}$ for the tube diameter d derived by Semenov [60] this line of reasoning was first used by Isambert and Maggs [79] to establish a scaling relation between plateau modulus and concentration. They argue that the macroscopic shear deformation is affinely passed down to the tube that are compressed or stretched depending on their orientation to the shear. The resulting change of the tube diameter causes a change in the deflection length L_d and with the help of (5.4) the resulting modulus scales as

$$G \propto c^{7/5} . \tag{5.5}$$

The same scaling was also obtained by other approaches, e.g. by a modified Onsager theory for confinement tubes [50]. Morse [80] has proposed different theoretical models for this scaling: a detailed microscopic description of the topological constraint imposed by neighboring polymers leads him to the prediction of a modulus scaling with $c^{7/5}$ and a quite different approach yields a scaling of $G \propto c^{4/3}$ derived by a self-consistent treatment of the network as an elastic continuum. Since these values are numerically quite close, a verification has not yet been possible with the accuracy of available experimental data.

It has to be kept in mind that all these theoretical approaches are implicitly build on the assumption that the tube contour deforms affinely with the macroscopic strain. This is evidently only a very coarse-grained description of the system's response. While stress relaxation by slow processes like reptation is obviously not relevant on the time scale of the plateau modulus, it is however possible that faster relaxation processes cause a tube contour that differs from the contour obtained by affine displacement. Our goal is to implement this relaxation processes by a free energy minimization in a numerical solution of the plateau modulus and investigate the quantitative and qualitative differences to the affine model. The detailed setup of this approach is discussed in the following section.

5.3 Numerical Solution

Our approach is to obtain a numerical solution of the partition sum (5.2) for a test polymer in a typical network of semiflexible polymers by averaging over all allowed configurations of neighboring polymers. The advantage of this approach is a microscopic description of the Hamiltonian. In contrast to the tube-model that only provides a coarse-grained description of the surrounding polymers, it accounts for the detailed interactions in a given realization of disorder. This permits to investigate the effect of local non-affine deformations of the encaged test polymer. Since the distribution of obstacle polymers around a given test filament is quite heterogeneous, it is expected that these non-affine deformations result in a lower global free energy. Our aim is to find this free energy minimum by a numeric minimization procedure.

5.3.1 Reduction to 2D

To reach this goal, we start by decomposing the transverse undulations of the test polymer into two independent components as previously described [108]. For one component the Hamiltonian thus simplifies to the description of a two-dimensional polymer in a plane surrounded by a certain number N_{obs} fluctuating point-like obstacles (see Figure 5.1). The point-like obstacles are subjected to a harmonic potential with strength γ around an equilibrium position p_j^0 with $j = 1, .., N_{\text{obs}}$. The parameters N_{obs} and γ can be chosen to self-consistently represent a network of a specific concentration [93]. Therefore the system is completely described by the two-dimensional contour $\mathrm{r}(s)$ of the test polymer and and N_{obs} two-dimensional vectors $\mathrm{p_j}$ describing the positions of the obstacles in the plane. The Hamiltonian thus reads $H = H^p + \sum_{j=1}^{N_{\text{obs}}} H_j^{\text{obs}}$ where H^p is the bending energy contribution

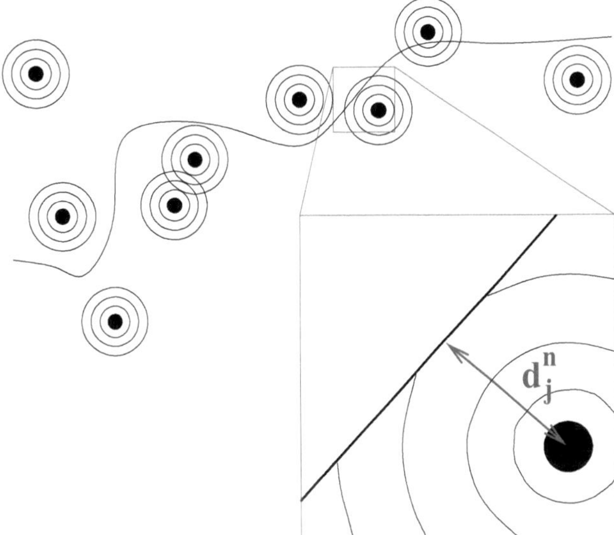

Figure 5.1: Fixed polymer in an array of fluctuating point obstacles (black points). The interaction between the polymer and the obstacles as the fluctuations of the j-th point obstacle are hindered by the polymer at a distance d_j^n (see inset).

r^0. A visualization of this mode and the mode $u_{k=3}$ is depicted in Figure 5.2. For a specific mode we can monitor the resulting free energy as a function of the mode amplitude A_k as exemplary shown in Figure 5.3. The result is a harmonic function $F(A_k) = \omega_k/2A_k^2 + F^0$ where ω_k can be determined from the plots for every mode. F^0 is the free energy of the contour $\mathrm{r}^0(s)$ with no modes excited and corresponds to the minimum in Fig. ??. The

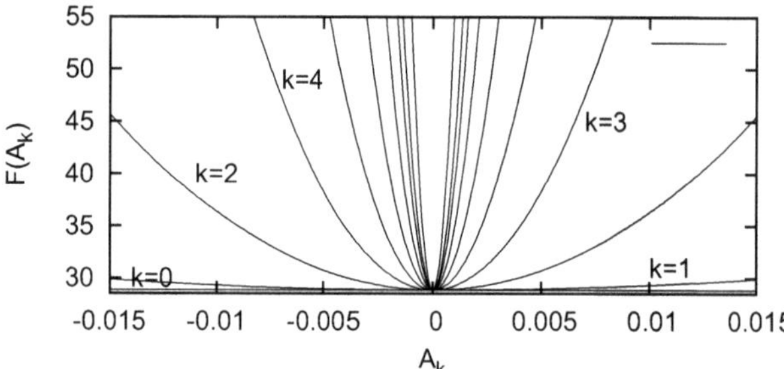

Figure 5.3: Increase of free energy with amplitude for different modes in a given random array of obstacles.

Hamiltonian for a certain contour that is in the representation (5.11) fully characterized by the set of coefficients $\{A_k\}$ is then given as

$$H(\{A_k\}) = \frac{1}{2}\sum_k \omega_k A_k^2 + F^0 \tag{5.12}$$

The desired partition function can thus easily be obtained by integration over the mode amplitudes

$$Z = \prod_k \int dA_k \exp[-H(\{A_k\})] = \exp[-\beta F^0] \prod_k \sqrt{\frac{2\pi}{\beta\omega_k}} \,. \tag{5.13}$$

This also allows for calculation of a measure for the system's free energy. The free energy cost of confinement could in principle be obtained by substracting for every mode the energy of a free polymer from the energy of the confined polymer. Since we are only interested in the deformation-induced free energy change, an arbitrary absolute value of the free energy is sufficient for our purpose. Technically, the product over the modes in Eq. (??) can be truncated as soon as the free energy change converges. This conversion is due to the fact that the steep increase of ω_k with k (compare Figure 5.3) is completely dominated by the bending energy contribution to the modes. Since this contribution is an intra-polymer property, it is independent of the test polymer's environment and does not change under deformation. This is explicitly shown for the tube model in Appendix D.

5.3.4 Shear Deformation

Having developed an approach to calculate the free energy of a fluctuating polymer in an array of fluctuating topological constraints, we can proceed to investigate the free energy change as a reaction to shear deformations. A simple example of such a deformation is a global shear deformation of a macroscopic sample. If the sample is an equilibrated network of semiflexible polymers, the result of the shear deformation will be a rise in free energy and consequently a force counter-acting the deformation. The system is thus perturbed by the deformation and brought to a non-equilibrium state which will immediately be followed by relaxation processes. These relaxation processes occur on very different time scales for the different length scales in the network. This is the reason for the frequency dependence of the modulus.

For very long time for instance, the network is able to completely relax the deformation stress by reptation thereby recovering the equilibrium value of free energy and resulting in a vanishing modulus. We can assume that at the time scale of the plateau modulus the encaged polymers have completely experienced their immediate surroundings but no large scale relaxation by network rearrangement has occurred. Thus in measuring the plateau modulus the tube has sufficient time to form before the deformation field changes again. This argument is the foundation for the assumption of affine displacement of the tube's contour and size. As it is unknown how exactly the macroscopic stress is passed on to the microscopic constituents of the network, it is commonly assumed that the deformation field follows the macroscopic stress on all length scales. Since at the time scale of the plateau modulus the tube is the relevant quantity, the tube centers or backbones are displaced affinely with the global shear. Translated to our two-dimensional plane of observation, this signifies that the tube contour $\mathrm{r}^0(s)$ and the centers of the obstacle tube p_j^0 are deformed affinely as depicted in Figure 5.4 (top). The free energy of this new configuration can be calculated as shown above. The modulus G_A to be determined from this resulting free energy change should be equivalent to the modulus obtained from any theoretical treatment that is based on the assumption of affine displacement.

If we take a closer look at the processes at tube formation, it becomes clear that the tube obtained by affine deformation is not necessarily a valid description of the physical reality. To this end we relax the assumption of affine displacement in such a way, that we now only let the obstacle fluctuation center p_j^0 deform affinely as illustrated in Figure 5.4 (center). If this new configuration of the obstacle array is taken for granted, we have to ask the question how the test polymer encaged by the obstacles reacts to this conformational change. Out of all possible deformations of the test polymer only the one with the lowest free energy will actually be realized. Obviously, this new tube contour will only in very few cases equal the tube contour obtained by affine displacement of the original contour. The resulting free energy of this deformation will thus be less or equal to the free energy obtained by affine deformation and we can therefore also state that the resulting modulus $G_{NA} \leq G_A$. Technically, the free energy difference of this non-affine deformation of the tube contour is obtained by displacing the p_j^0 affinely with the macroscopic stress (see E) and then applying again the free energy minimization as explained above to find the new

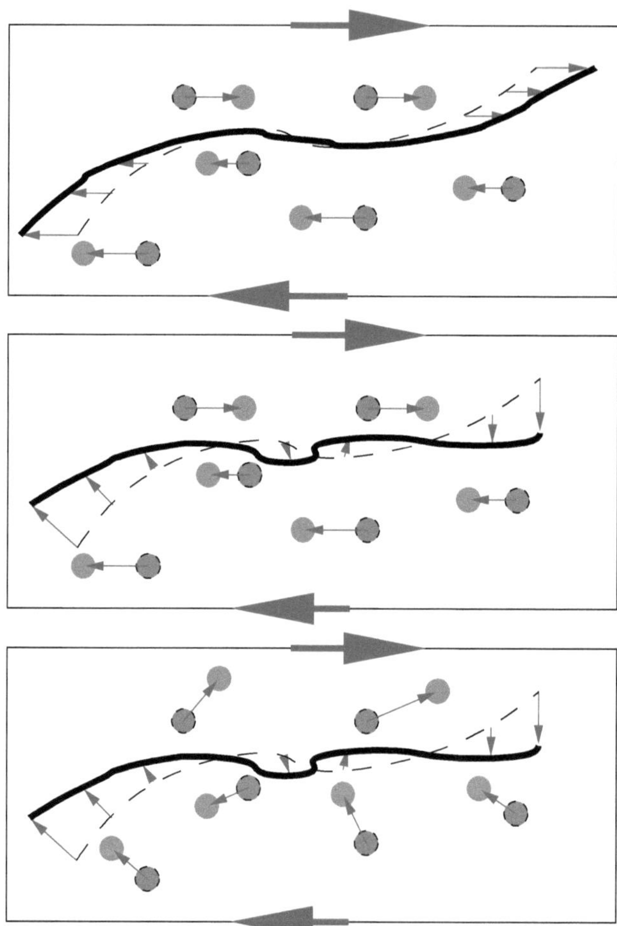

Figure 5.4: Different levels of affinity in shear deformations. (top) Obstacle fluctuation centers and tube backbone follow the macroscopic shear deformation (large red arrow) affinely. (middle) While the obstacle fluctuation centers follow the macroscopic shear deformation affinely, the tube backbone can deviate from the affine deformation field in order to minimize the global free energy. (bottom) Also the obstacle point can react by non-affine deformation to the macroscopic shear.

tube contour.

Of course, also the fact that we deform the obstacle points affinely implies an assumption. In the actual physical system the obstacle points and with it the tube centers of the neighboring polymers are free to change their position in order to reach a global state of lower free energy (see Figure 5.4 (bottom). As these neighboring polymers however couple to other polymers outside our plane of observation, the incorporation of this feature would be tantamount to a minimization in all degrees of freedom of the network. This is obviously out of range of a numerical solution. The resulting modulus of a complete free energy minimization is the modulus G that would be observed in experiments. The modulus determined by our approach constitutes an upper bound for the experimental values and thus $G \leq G_{NA}$.

5.4 Results

In the previous section we presented an approximative numerical solution to the problem of finding the free energy change under shear deformation of a single probe polymer. This is obviously a quantity that can not be observed experimentally, but it can serve as the basis for the calculation of the macroscopic plateau modulus. To obtain this observable we have to add up the contributions from all polymers in the network under consideration. One single specific polymer is described in terms of the two dimensional plane of observation whose orientation in space is described by a set of angles (θ, ϕ, ψ) as described in Appendix E. Consequently, we have to perform an average over these isotropically distributed angles and furthermore we have to average over the quenched disorder that is generated by the different configurations of point-like obstacles in the observation plane. With the shear parameter Γ this procedure finally returns an average free energy function $\Delta F(\Gamma) = g\Gamma^2/2$ from which the macroscopic modulus is obtained as $G = 2\nu g$. The factor 2 stems from the fact that every polymer is described by two planes of observation corresponding to the two components of transverse fluctuation.

For a single polymer in a plane and one specific realization of obstacle disorder the resulting free energy function is exemplary shown in Figure 5.5 (top). Since the absolute value of the free energy differs strongly with the actual obstacle configuration all plots have been rescaled to the free energy value at $\Gamma = 0$. Obviously, for a single polymer in different specific realizations of obstacle disorder and different orientations of the plane of observation to the applied shear the resulting form of the free energy is highly variable. Furthermore, the free energy minimum is in general not at the point of zero shear. This feature however, should of course be fulfilled for the free energy function that is obtained by summing up all constituents in an macroscopic sample at equilibrium. We chose to use this requirement as a verification for an sufficient sampling over disorder. The location of the accumulated free energy minimum initially strongly oscillates with the number of samples but finally converges to $\Gamma = 0$. For every data point we average over a sufficient number of disorder samples until this criterion is fulfilled.

An example of the resulting averaged free energy function is shown in Figure 5.6.

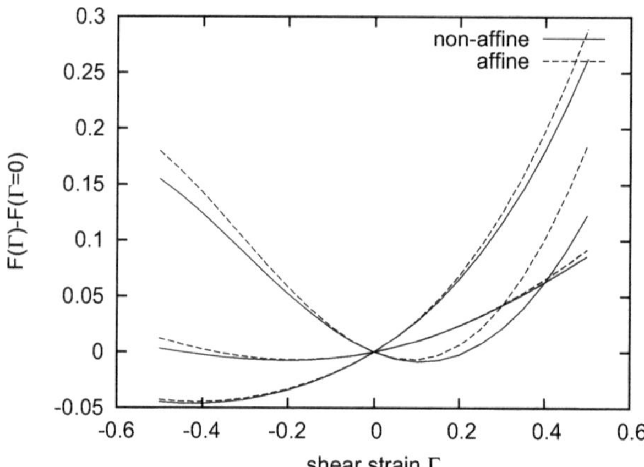

Figure 5.5: Free energy change with shear Γ for three different realizations of a test polymer in a network.

As expected the minimum is at zero shear where the system is at equilibrium. At the application of small shear the free energy rises in a harmonic fashion which would entail a linear restoring force in an experimental measurement. Furthermore, it can be seen that the free energy obtained in the affine approximation is always above the non-affine free energy that was obtained by the minimization procedure explained above. We determine the modulus by an harmonic fit at small shear strains. At higher strains however, the free energy function is no longer faithfully described by this fit, but features a stronger slope. This signifies the onset of non-linear forces.

5.4.1 Affine vs. Non-Affine

We determined the resulting plateau moduli as a function of different system parameters. Figure 5.7 illustrates the scaling of the modulus with polymer concentration c. Both the affine and the non-affine modulus show good agreement to a 7/5 power law with concentration. The non-affine modulus is considerably below the affine modulus. This confirms the initial assumption that a deformation field that assumes affine displacement on all length scales indeed over estimates the system's response. The non-affine deformation that is obtained by permitting the encaged polymer to find its tube of minimal free energy leads to a lower modulus. Comparing the moduli from our affine calculation with the prediction for the absolute plateau modulus by Morse's "Binary Collision Approximation" [80], shows sound consistency. In the realm of the restriction of affine displacement, our work can be seen as a numerical confirmation. However, the moduli obtained by experimental

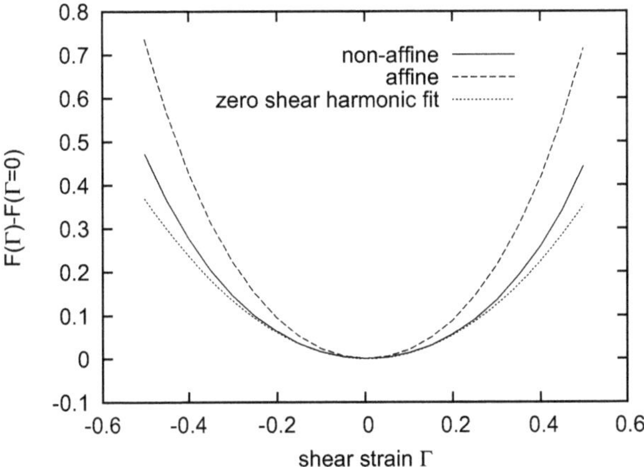

Figure 5.6: Free energy change obtained by averaging over quenched disorder and orientation where the lower free energy curve is due to energy minimized non-affine deformations. Harmonic fits are only a valid approximation in the linear regime at small shears.

measurements [50] are considerably lower and prove that the physical reality is closer to a non-affine deformation field. Of course, the detailed nature of this field is not accessible to experiments but our data suggests that the proposed model of an affine displacement of neighbors combined with an non-affine displacement of the tube is an appropriate approximation. The non-affine moduli obtained by this approach only show slight overestimations of the experimental results and it can be argued that this is due to possible additional non-affinities in the obstacle displacement.

5.4.2 Scaling with Persistence

Finally, we determined the scaling of the plateau modulus with persistence length l_p. The decrease of the modulus with increasing polymer stiffness is depicted in Figure 5.8 and shows good agreement with a power law of $G \propto l_p^{-1/5}$. The persistence dependence represents a sensible method to discriminate between competing models of viscoelasticity. Since the value of the concentration scaling exponent $4/3$ predicted by an effective medium approach [80] is numerically quite close to the exponent $7/5$ predicted by most other theories, experimental accuracy does not allow for a verification. The difference of the two concepts in the persistence length scaling exponents is considerably larger: $-1/3$ versus $-1/5$. Recently, data was presented in an experimental work [105] that claimed to have verified a scaling exponent of $-1/3$. However, the authors only provided measurements at two different values of persistence length. We present data over a range of persistence lengths global energy. We observed non-affine moduli that agree well with experimental data. For comparison we also calculated affine moduli that are considerably higher and coincide with the results of existing theoretical predictions. Furthermore, we clearly confirm a scaling of the plateau modulus with persistence length and concentration as $G \propto c^{7/5} l_p^{-1/5}$ and are able to resolve the dispute between different theoretical models. Our results prove that shear deformation of networks of entangled polymers has to be described in a non-affine picture and that affine theories systematically overestimate the mechanical response. The presented approach provides a numerical solution to evaluate complex partition sums and has a wide applicability to rheology of polymer networks. Future applications can e.g. investigation of non-linear shear.

Chapter 6

Summary and Outlook

The focus of this work was the investigation of biopolymer networks. These networks form complex materials and play a vital role as one of the cytoskeleton's main constituents. Since the cytoskeleton acts as the structural element of the eucaryotic cell, its investigation is crucial for the understanding of a variety of dynamical and mechanical properties in cell biology. In particular, we have studied entangled networks of semiflexible polymers, where the only interaction between polymers is their mutual uncrossability.

Due to this topological interaction, filaments can effortlessly slide past each other but are not allowed to cross. By reptation through thermal motion every polymer can thus realize any conceivable configuration on long time scales. On intermediate time scales however, the accessible configuration space is restricted by surrounding filaments and can be described by a tube. The tube concept permits to resort to a tractable single polymer description where the effect of all neighboring filaments is condensed into a virtual tube.

A central part of this work is devoted to the precise characterization of the confinement tube in terms of size and form. While the tube model has so far been a successful qualitative concept, we have developed an analytical and quantitative theory for the absolute value of the tube diameter $L_\perp$ as a function of the polymers' persistence length l_p and mesh size ξ of the network. To leading order we find $L_\perp = 0.31\ \xi^{6/5} l_\mathrm{p}^{-1/5}$, which is consistent with known asymptotic scaling laws. Additionally, our theory provides finite length corrections that can account for effects of polydispersity. We have supported our analytical studies by extensive computer simulations and could also verify the harmonic form of the tube potential. Furthermore, we go beyond the description of the tube in terms of its average size and study its conformation. To this end we have developed a simulational approach to analyze curvature distributions of confinement tubes and single confined filaments where we have implemented a model that closely mimics real polymer dynamics. In accordance with experimental measurements we unexpectedly find distribution functions that feature distinctive differences from free polymers even in the absence of excluded volume. Extensive simulations allow to attribute these features to entropic trapping in network void spaces and we are able to show that these transient non-equilibrium are a generic effect observed at all time scales below large scale rearrangement as relevant for most biological processes and experimental observations.

Finally, we have investigated non-equilibrium properties of polymer networks, where our aim was a quantitative theory of the system's mechanical response to shear deformations. Specifically, we have developed a description of the plateau modulus based on the microscopic intra-polymer interactions. While existing theories provide different qualitative predictions for this observable, we have presented an approach that allows for a direct numerical computation of the system's free energy. Contrary to prior descriptions that assume an affine deformation field on all length scales, our model permits for the first time to obtain local non-affine deformations by a global minimization of free energy. We show the importance of these non-affine deformations for the absolute value of the modulus and compare our results with experimental measurements. It is found that non-affine deformations are essential for a correct description and that the assumption of affine displacement considerably overestimates the material response. Furthermore, we provide an independent and precise confirmation of the scaling $G \propto c^{7/5} l_p^{-1/5}$ of the modulus G with polymer concentration c and persistence length l_p.

In summary, the results presented in this thesis fundamentally deepen the understanding of entangled networks of semiflexible polymers. Existing concepts are expanded by quantitative predictions as for example the absolute value of the tube diameter, its length corrections or the absolute value of the modulus. Additionally, new concepts like transient distribution functions and non-affine deformations are introduced and their significance is emphasized.

As an outlook, several extensions and continuations of our work can be imagined. Additionally to the absolute value of the average tube diameter we have provided simulation data of the distribution of tube diameters in the network. To gain a deeper understanding of the heterogeneities in the network it would be promising to develop a quantitative theory for these distributions. This is also expected to be closely connected to the phenomenon of entropic trapping. While we have gained a qualitative understanding of this property, a more detailed analysis providing scaling laws or even quantitative predictions would be desirable. The elaboration of our work on tube and polymer conformations could aim at an improved description of reptation phenomena. This would also allow for experimental verification and contribute to the ongoing puzzle of transport in disordered environments like the cell's cytoplasm. Concerning the mechanical properties, our numerical approach provides an excellent tool to study microscopic non-affine deformation fields that are not accessible to experiments and can also be extended non the investigation of non-linear shear.

Appendix A

Rigid Rod Statistics I

To relate the polymer concentration of a network to the obstacle density per unit area in the simulation, we calculate the number of randomly distributed and oriented stiff rods per unit volume that intersect with a unit plane. Every intersecting rod will be described by its polar and azimuth angle relative to the unit plane, the point of intersection and the distance between center of mass and intersection point. Because of rotational symmetry the problem is independent of the azimuth angle and because of uniform density it is independent of one of the coordinates of the intersection point. Hence, the problem is equivalent in two dimensions to the number of rods per unit area that intersect a unit line (see Fig. A.1). The plane contains ν rigid rods per unit area with random orientation α and center of mass position. The number of intersections ρ_{MC} with the unit line (bold dashed) is computed by parameterizing the center of mass (C) by the coordinate z of the intersection point (P) with the unit line, the distance s between C and P and the angle α. ρ_{MC} is then obtained as the integral over all possibly intersecting rods:

$$\rho_{\mathrm{MC}} = \frac{2\nu}{2\pi} \int dr^2 \int_0^{\pi} d\alpha \; , \tag{A.1}$$

where the factor 2 accounts for the fact that any rod configuration can be realized by two angles α since the rods have no direction. The integration area has to be chosen appropriately to only include intersecting rods. As $r = (\sin(\alpha)s, z - \cos(\alpha)s)$ the Jacobian determinant of the coordinate transformation to integration variables is $\partial r/\partial(z, s) = \sin(\alpha)$ and the integral evolves to

$$\rho_{\mathrm{MC}} = \frac{\nu}{\pi} \int_{-1/2}^{1/2} dz \int_{-L/2}^{L/2} ds \int_0^{\pi} d\alpha \sin(\alpha) = \frac{2}{\pi}\nu L \; . \tag{A.2}$$

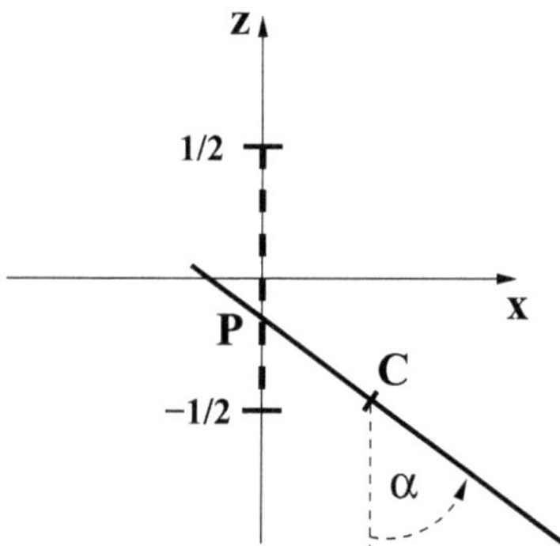

Figure A.1: Sketch of a rod with center of mass (C) and orientation α intersecting a arbitrary unit line (dashed) to illustrate the calculation of the number of intersecting rods ρ_{MC}.

Appendix B

Rigid Rod Statistics II

We derive the radial density of obstacles that effectively hinder the test rods fluctuations. To this end we consider the test rod to be aligned along the z-axis without loss of generality (see Fig. B.1). As criterion for effective obstruction of transverse fluctuations between test rod and obstacle, we demand that the line connecting their points of closest approach ($\overline{PO}$) is orthogonal to both polymers. As a projection of the obstacle to the plane spanned by the test rod and $\overline{PO}_\perp$ (dashed) recovers the setup discussed in Appendix A, the coordinates of the center of mass (C) can readily be extended to three dimensions by the radial distance R and the angle β to $r = (\sin(\alpha)s\sin(\beta) - \cos(\beta)R, \sin(\alpha)s\cos(\beta) + \sin(\alpha)R, x - \cos(\alpha)s)$ and with the Jacobian determinant $|\partial r/\partial(x, s, R)| = \sin(\alpha)$ the integration gives

$$\begin{aligned} \rho(R) &= \frac{2\nu}{4\pi}\int_{-\frac{L}{2}}^{\frac{L}{2}} dz \int_{-L/2}^{L/2} ds \int_0^{\pi} d\alpha \int_0^{2\pi} d\beta \sin^2(\alpha) \\ &= \frac{\pi}{2}\nu L^2 . \end{aligned} \tag{B.1}$$

Consequently, the density seen by an stiff segment of length $\bar{L}$ in the IRM will be

$$\rho = \frac{\bar{L}}{L}\frac{\pi}{2}\nu L^2 = \frac{\pi}{2}\nu L\bar{L} . \tag{B.2}$$

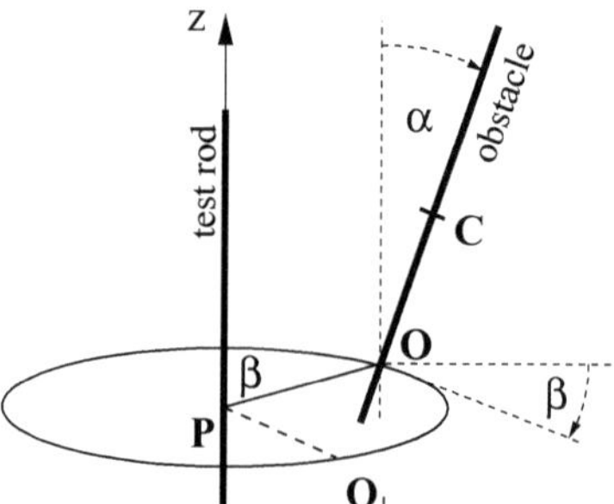

Figure B.1: Two rods with minimal distance R are considered to be mutually interacting only if their line of closest approach ($\overline{OP}$) is orthogonal to both. $O_\perp$ only serves to illustrate the analogy to the two-dimensional setup (see text).

Appendix C

Mode analysis of polymer and tube

Using the dimensionless arc-length $\tilde{s} = s/L$, polymer conformation $\tilde{y}(\tilde{s}) = y(\tilde{s}L)/L$, tube center $\tilde{y}^0(\tilde{s}) = y^0(\tilde{s}L)/L$ and persistence length $\eta = l_\mathrm{p}/L$ the Hamiltonian (3.1) can be written for one transverse coordinate as

$$\beta H = \frac{\eta}{2} \int_0^1 d\tilde{s} \left[(\partial_{\tilde{s}}^2 \tilde{y})^2 + l_\mathrm{d}^{-4} (\tilde{y} - \tilde{y}^0)^2 \right] . \tag{C.1}$$

For free boundary conditions the Hamiltonian is diagonalized by an orthonormal set of eigenfunctions $\psi_k \approx \sin(q_k\sigma + \phi_k)$ with $q_k \approx \pi(k+1/2)$ [111]. Expanding both polymer and tube center in modes as $\tilde{y}(\sigma) = \sum_k y_k \psi_k$ and $\tilde{y}^0(\sigma) = \sum_k y_k^0 \psi_k$ and using the orthogonality of the eigenfunctions allows one to write the Hamiltonian in the suggestive form

$$\begin{aligned} \beta H &= \frac{\eta}{2} \sum_k (q_k^4 + l_\mathrm{d}^{-4}) \left(y_k - \frac{y_k^0}{1 + (q_k l_\mathrm{d})^4} \right)^2 \\ &\quad + \frac{q_k^4 y_k^{0\,2}}{1 + (q_k l_\mathrm{d})^4} . \end{aligned} \tag{C.2}$$

By comparison to (C.1) one can read of the minimum of the confinement potential, i.e. the average tube center:

$$\overline{y_k} = y_k^0 / (1 + q_k^4 l_\mathrm{d}^4) . \tag{C.3}$$

Additionally, this allows to write the complete transverse fluctuations as a sum over the inverse confinement strength of all modes:

$$L_\perp^2 := \frac{1}{L} \int_0^L \overline{ds (x(s) - \overline{x(s)})^2} = \frac{L^2}{\eta} \sum_k \frac{1}{q_k^4 + l_\mathrm{d}^{-4}} . \tag{C.4}$$

For the dimensionless function $h(l_\mathrm{d})$ relating the tube diameter to the deflection length follows:

$$h(ld) = \sum_k \frac{1}{q_k^4 + l_\mathrm{d}^{-4}} . \tag{C.5}$$

Appendix D

Mode Representation of Free Energy

To explain the vanishing contribution of higher modes to the free energy change, we resort once again to a description of a confined polymer in the tube model and first compute the confinement free energy. In the weakly-bending rod approximation the Hamiltonian is written as $H(\gamma, \kappa) = \int_0^L ds[\kappa/2(\partial_s^2 r_\perp(s))^2 + \gamma/2 r_\perp^2]$ with tube potential strength γ. The Hamiltonian for a free polymer is obtained by simply dropping the last term. By using the eigenmode representation from Eq. (5.11) it reduces to

$$H(A_k, \gamma, \kappa) = \frac{L}{2} \sum_k \frac{\kappa}{2} \frac{k^4}{L^4} A_k^2 + \frac{\gamma}{2} A_k^2 \,. \tag{D.1}$$

By integrating over the amplitudes the free energy F^c of the confined polymer is thus obtained as

$$F^c = -k_{\mathrm{B}} T \ln \prod_{k=0}^{k_c ut} \sqrt{\frac{\pi}{L(\frac{\kappa k^4}{L^4} + \gamma)}} \tag{D.2}$$

and the free energy F^f of the corresponding free polymer as

$$F^f = -k_{\mathrm{B}} T \ln \prod_{k=0}^{k_c ut} \sqrt{\frac{\pi}{L(\frac{\kappa k^4}{L^4})}} \,. \tag{D.3}$$

Consequently, the free energy cost of confinement is the difference

$$\Delta F = F^c - F^f = \sum_k \ln \sqrt{\frac{\frac{\kappa k^4}{L^4} + \gamma}{\frac{\kappa k^4}{L^4}}} \tag{D.4}$$

that obviously converges with increasing k. Let us assume this is the confinement free energy at zero shear $\Delta F(\Gamma = 0)$. If we now apply a shear deformation, we are interested in the free energy change $\Delta F(\Gamma) - \Delta F(\Gamma = 0)$ to compute the mechanical response. Since the deformation only affects the tube potential, the free energy in Eq. (D.4) is only affected via γ. As the γ term is quickly dominated by the k^4 term with increasing mode number k, we can expect that the higher modes do not contribute significantly to the free energy change $\Delta F(\Gamma) - \Delta F(\Gamma = 0)$ upon shear deformation.

Appendix E

Shear Deformation

We work with two coordinate systems: a three dimensional real space system and a two dimensional system in the plane of observation, where the origin of both systems is one end of the initial test polymer. The orientation of the end-to-end vector R of an arbitrary test polymer in the three dimensional space is isotropically distributed. As depicted in Fig. E.1 it is described by the two angles θ and ϕ. To define a plane of fluctuations we need one additional angle ψ. This plane is spanned by the vectors R and S with $\mathrm{S} \perp \mathrm{R}$. For $\psi = 0$ the vector S is obtained by applying the same transformation to an vector parallel to the z-axis, that is needed to transform an vector parallel to the x-axis to R. Other values of ψ are obtained by an rotation around the axis R. Graphically it is helpful to picture the observation plane as the plane that is obtained by applying two transformations to the x-z-plane: first a rotation around the z-axis by ϕ and then a rotation around the axis R by ψ. If R and S are normalized they correspond to the x and y-axis in the plane of observation.

If we now apply a macroscopic shear deformation $\mathcal{T}(\Gamma)$ to the three dimensional system the obstacle points and the test polymer deform according to the action of the transformation $\mathcal{T}$ on their real space coordinates. In general this signifies that they leave the plane of observation spanned by R and S. It is however self-evident that also the test polymer's fluctuations are subjected to the shear and therefore also the plane of observation transforms according to $\mathcal{T}$. This is tantamount to a transformation of the vectors R and S and leaves the transformed obstacles points in the new plane of observation. We obtain the new plane coordinates in terms of the transformed unit vectors R' and S' that correspond again to x and y-axis. The former is obtained as $\mathrm{R}' = \mathcal{T}(\Gamma)\mathrm{R}$ and the latter is constructed as the component of $\mathcal{T}(\Gamma)\mathrm{S}$ that is orthogonal to R'.

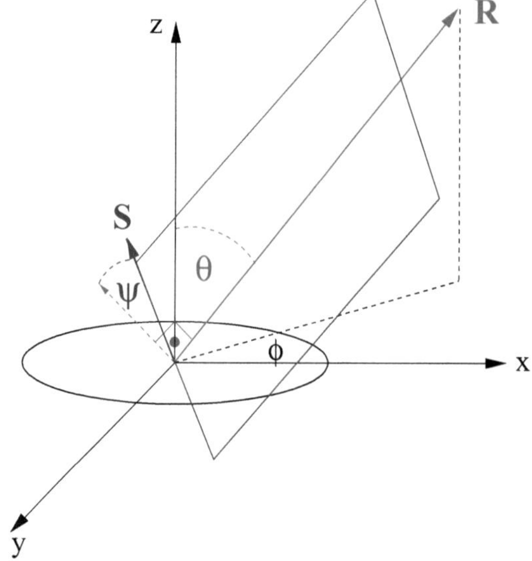

Figure E.1: Schematic illustration of the plane of observation.

Bibliography

[1] T. Hiiragi and D. Solter, First cleavage plane of the mouse egg is not predetermined but defined by the topology of the two apposing pronuclei, Nature 430, 360 (2004).

[2] M. F. Carlier, Actin: Protein structure and filament dynamics, J. Bio. Chem. 266, 1 (1991).

[3] T. Pollard, Introduction to actin and actin-binding proteins. In Guidebook to the Cytoskeletal and Motor Proteins, Oxford University Press, 1999.

[4] T. M. Svitkina and G. G. Borisy, Arp2/3 complex and actin depolymerizing factor/cofilin in dendritic organization and treadmilling of actin filament array in lamellipodia, J. Cell Bio. 145, 1009 (1999).

[5] T. D. Pollard and G. G. Borisy, Cellular motility driven by assembly and disassembly of actin filamentsv, Cell 112 (2003).

[6] U. Euteneuer and M. Schliwa, Persistent, directional motility of cells and cytoplasmic fragments in the absence of microtubules, Nature 310, 58 (1984).

[7] H. E. Huxley, The mechanism of muscular contraction, Science 164, 1356 (1969).

[8] I. Rayment, H. Holden, M. Whittaker, C. Yohn, M. Lorenz, K. Holmes, and R. Milligan, Structure of the actin-myosin complex and its implications for muscle contraction, Science 261, 58 (1993).

[9] A. Desai and T. J. Mitchison, Microtubule polymerization dynamics, Ann. Rev. Cell Devel. Bio. 13, 83 (1997).

[10] I. R. Gibbons and A. J. Rowe, Dynein: A protein with adenosine triphosphatase activity from cilia, Science 149, 424 (1965).

[11] R. D. Vale, T. S. Reese, and M. P. Sheetz, Identification of a novel force-generating protein, kinesin, involved in microtubule-based motility, Cell 42, 39 (1985).

[12] R. D. Vale and R. A. Milligan, The way things move: Looking under the hood of molecular motor proteins, Science 288, 88 (2000).

[13] M. Schliwa and G. Woehlke, Molecular motors, Nature 422, 759 (2003).

[14] M. Kirschner and T. Mitchinson, Beyond self-assembly: from microtubules to morphogenesis, Cell 45, 329 (1986).

[15] T. J. Mitchison and E. D. Salmon, Mitosis: a history of division, Nature Cell Bio. 3, E17 (2001).

[16] I. R. Gibbons, Cilia and flagella of eukaryotes, J. Cell Bio. 91, 107 (1981).

[17] G. Ruben, A. Alonso, I. Grundke-Iqbal, and K. Iqbal, Taxol stabilized rat brain microtubules with microtubule-associated proteins (maps) freeze-dried and vertically platinum-carbon (pt-c) replicated: New ultra-high resolution images for evaluating the relationship of maps to microtubules, Neuroscience-Net 1, 10002 (1996).

[18] H. Felgner, R. Frank, and M. Schliwa, Flexural rigidity of microtubules measured with the use of optical tweezers, J. Cell Sci. 109, 509 (1996).

[19] F. Pampaloni, G. Lattanzi, A. Jonas, T. Surrey, E. Frey, and E.-L. Florin, Thermal fluctuations of grafted microtubules provide evidence of a length-dependent persistence length, Proc. Nat. Acad. Sci. 103, 10248 (2006).

[20] H. Isambert, P. Venier, A. Maggs, A. Fattoum, R. Kassab, D. Pantaloni, and M. Carlier, Flexibility of actin filaments derived from thermal fluctuations. effect of bound nucleotide, phalloidin, and muscle regulatory proteins, J. Biol. Chem. 270, 11437 (1995).

[21] A. Ott, M. Magnasco, A. Simon, and A. Libchaber, Measurement of the persistence length of polymerized actin using fluorescence microscopy, Phys. Rev. E 48, R1642 (1993).

[22] F. Gittes, B. Mickey, J. Nettleton, and J. Howard, Flexural rigidity of microtubules and actin filaments measured from thermal fluctuations in shape, J. Cell Biol. 120, 923 (1993).

[23] P. Janmey, S. Hvidt, J. Kas, D. Lerche, A. Maggs, E. Sackmann, M. Schliwa, and T. Stossel, The mechanical properties of actin gels. elastic modulus and filament motions, J. Bio. Chem. 269, 32503 (1994).

[24] O. Kratky and G. Porod, Röntgenuntersuchungen gelöster fadenmoleküle, Rec. Trav. Chim. 68, 1106 (1949).

[25] N. Saito, K. Takahashi, and Y. Yunoki, The statistical mechanical theory of stiff chains, J. Phys. Soc. Jpn. 22, 219 (1967).

[26] H. Yamakawa and M. Fujii, Wormlike chains near the rod limit: Path integral in the wkb approximation, J. Chem. Phys. 59, 6641 (1973).

[27] H. E. Daniels, The statistical theory of stiff chains, Proc. Roy. Soc. Edingburgh 63, 290 (1952).

[28] J. Wilhelm and E. Frey, Radial distribution function of semiflexible polymers, Phys. Rev. Lett. 77, 2581 (1996).

[29] K. Kroy and E. Frey, Force-extension relation and plateau modulus for wormlike chains, Phys. Rev. Lett. 77, 306 (1996).

[30] K. Binder, Monte Carlo and molecular dynamics simulations in polymer science, Oxford University Press, 1995.

[31] D. Frenkel and B. Smit, Understanding Molecular Simulations, Academic Press, San Diego, 1996.

[32] F. C. MacKintosh, J. Käs, and P. A. Janmey, Elasticity of semiflexible biopolymer networks, Phys. Rev. Lett. 75, 4425 (1995).

[33] O. Lieleg, M. M. A. E. Claessens, C. Heussinger, E. Frey, and A. R. Bausch, Mechanics of bundled semiflexible polymer networks, Phys. Rev. Lett. 99, 088102 (2007).

[34] C. Heussinger, M. Bathe, and E. Frey, Statistical mechanics of semiflexible bundles of wormlike polymer chains, Phys. Rev. Lett. 99, 048101 (2007).

[35] M. Bathe, C. Heussinger, M. M. A. E. Claessens, A. R. Bausch, and E. Frey, Cytoskeletal bundle mechanics, Biophys. J. 94, 2955 (2008).

[36] D. A. Head, A. J. Levine, and F. C. MacKintosh, Deformation of cross-linked semiflexible polymer networks, Phys. Rev. Lett. 91, 108102 (2003).

[37] J. Wilhelm and E. Frey, Elasticity of stiff polymer networks, Phys. Rev. Lett. 91, 108103 (2003).

[38] F. J. Nedelec, T. Surrey, A. C. Maggs, and S. Leibler, Self-organization of microtubules andmotors, Nature 389, 305 (1997).

[39] D. Humphrey, C. Duggan, D. Saha, D. Smith, and J. Käs, Active fluidization of polymer networks through molecular motors, Nature 416, 413 (2002).

[40] J. Uhde, M. Keller, E. Sackmann, A. Parmeggiani, and E. Frey, Internal motility in stiffening actin-myosin networks, Phys. Rev. Lett. 93, 268101 (2004).

[41] D. Pantaloni, C. L. Clainche, and M. F. Carlier, Mechanism of actin-based motility, Science 292, 1502 (2001).

[42] Y. Marcy, J. Prost, M. F. Carlier, and C. Sykes, Forces generated during actin-based propulsion: A direct measurement by micromanipulation, Proc. Nat. Acad. Sci. USA 101, 5992 (2004).

[43] J. M. Schurr and K. S. Schmitz, Dynamic light scattering studies of biopolymers: Effects of charge, shape, and flexibility, Ann. Rev. Phys. Chem. 37, 271 (1986).

[44] K. Kroy and E. Frey, Dynamic scattering from solutions of semiflexible polymersi, Phys. Rev. E 55, 3092 (1997).

[45] M. Hohenadl, T. Storz, H. Kirpal, K. Kroy, and R. Merkel, Desmin filaments studied by quasi-elastic light scattering, Biophys. J. 77, 1299 (19999).

[46] T. T. Perkins, D. E. Smith, and S. Chu, Direct observation of tube-like motion of a single polymer chain, Science 264, 819 (1994).

[47] J. Käs, H. Strey, and E. Sackmann, Direct imaging of reptation for semiflexible actin filaments, Nature 368, 226 (94).

[48] J. Käs, H. Strey, J. X. Tang, D. Finger, R. Ezzell, E. Sackmannt, and P. A. Janmey, F-actin, a model polymer for semiflexible chains in dilute, semidilute, and liquid crystalline solutions, Biophys. J. 70, 609 (1996).

[49] O. Mueller, H. E. Gaub, M. Baermann, and E. Sackmann, Viscoelastic moduli of sterically and chemically cross-linked actin networks in the dilute to semidilute regime: measurements by oscillating disk rheometer, Macromolecules 24, 3111 (1991).

[50] B. Hinner, M. Tempel, E. Sackmann, K. Kroy, and E. Frey, Entanglement, elasticity, and viscous relaxation of actin solutions, Phys. Rev. Lett. 81, 2614 (1998).

[51] F. Ziemann, J. Radler, and E. Sackmann, Local measurements of viscoelastic moduli of entangled actin networks using an oscillating magnetic bead micro-rheometer, Biophys. J. 66, 2210 (1994).

[52] J. C. C. andM. T. Valentine, E. R. Weeks, T. Gisler, P. D. Kaplan, A. G. Yodh, and D. A. Weitz, Two-point microrheology of inhomogeneous soft materials, Phys. Rev. Lett. 85, 888 (2000).

[53] M. Gardel, M. Valentine, J. Crocker, A. Bausch, and D.A.Weitz, Microrheology of entangled f-actin solutions, Phys. Rev. Lett. 91, 158302 (2003).

[54] P. G. de Gennes, Scaling Concepts in Polymer Physics, Cornell University Press, Ithaca, NY, 1979.

[55] M. Doi and S. F. Edwards, The Theory of Polymer Dynamics, Clarendon Press, Oxford, 1986.

[56] K. M. Schmoller, O. Lieleg, and A. R. Bausche, Cross-linking molecules modify composite actin networks independently, Phys. Rev. Lett. 101, 118102 (2008).

[57] M. Romanowska, H. Hinsch, N. Kirchgessner, M. Giesen, M. Degawa, B. Hoffmann, E.Frey, and R.Merkel, Direct observation of the tube model in f-actin solutions: Tube dimensions and curvatures, Euro. Phys. Lett. (2009).

[58] T. Odijk, On the statistics and dynamics of confined or entangled stiff polymers, Macromolecules 16, 1340 (1983).

[59] T. Burkhardt, Free energy of a semiflexible polymer confined along an axis, J. Phys. A. 28, 629 (1995).

[60] A. N. Semenov, Dynamics of concentrated solutions of rigid-chain polymers part 1. - brownian motion of persistent macromolecules in isotropic solution, J. Chem. Soc. Faraday. Trans. 82, 317 (1986).

[61] M. Doi and S. F. Edwards, Dynamics of rod-like macromolecules in concentrated solution., J. Chem. Soc. Faraday Trans. II 74, 1789 (1978).

[62] R. Götter, K. Kroy, E. Frey, M. Bärmann, and E. Sackmann, Dynamic light scattering from semidilute actin solutions: A study of hydrodynamic screening, filament bending stiffness, and the effect of tropomyosin/troponin-binding, Macromolecules 29, 30 (1996).

[63] S. F. Edwards and K. E. Evans, Dynamics of highly entangled rod-like molecules, J. Chem. Soc. Faraday Trans. 2 78, 113 (1982).

[64] I. Teraoka, N. Ookubo, and R. Hayakawa, Molecular theory on the entanglement effect of rodlike polymers, Phys. Rev. Lett. 55, 2712 (1985).

[65] C. Semmrich, T. Storz, J. Glaser, R. Merkel, A. R. Bausch, and K. Kroy, Glass transition and rheological redundancy in f-actin solutions, Proc. Natl. Acad. Sci. USA 104, 20199 (2007).

[66] M. A. Dichtl and E. Sackmann, Microrheometry of semiflexible actin networks through enforced single-filament reptation: Frictional coupling and heterogeneities in entangled networks, Proc. Natl. Acad. Sci. USA 99, 6533 (2002).

[67] M. Keller, R. Tharmann, M. A. Dichtl, A. R. Bausch, and E. Sackmann, Slow filament dynamics and viscoelasticity in entangled and active actin networks, Phil. Trans. Roy. Soc. London A 361, 699 (2003).

[68] J. Xu, A. Palmer, and D. Wirtz, Rheology and microrheology of semiflexible polymer solutions: Actin filament networks, Macromolecules 31, 6486 (1998).

[69] D. C. Morse, Viscoelasticity of concentrated isotropic solutions of semiflexible polymers. 2. linear response, Macromolecules 31, 7044 (1998).

[70] F. G. Schmidt, B. Hinner, and E. Sackmann, Microrheometry underestimates the values of the viscoelastic moduli in measurements on f-actin solutions compared to macrorheometry, Phys. Rev. Lett. 61, 5646 (2000).

[71] J. Liu, M. L. Gardel, K. Kroy, E. Frey, B. D. Hoffman, J. C. Crocker, A. R. Bausch, and D. A. Weitz, Microrheology probes length scale dependent rheology, Phys. Rev. Lett. 96, 118104 (2006).

[72] F. Gittes and M. MacKintosh, Dynamic shear modulus of a semiflexible polymer network, Phys. Rev. E 58 (1998).

[73] D. C. Morse, Viscoelasticity of tightly entangled solutions of semiflexible polymers, Phys. Rev. E 58, 1237 (1998).

[74] F. Amblard, A. C. Maggs, B. Yurke, A. N. Pargellis, and S. Leibler, Subdiffusion and anomalous local viscoelasticity in actin networks, Phys. Rev. Lett. 77, 4470 (1996).

[75] F. Gittes, B. Schnurr, P. Olmsted, F. MacKintosh, and C. Schmidt, Microscopic viscoelasticity: Shear moduli of soft materials determined from thermal fluctuations, Phys. Rev. Lett. 79, 3286 (1997).

[76] B. Schnurr, F. Gittes, F. MacKintosh, and C. Schmidt, Determining microscopic viscoelasticity in flexible and semiflexible polymer networks from thermal fluctuations, Macromolecules 30, 7781 (1997).

[77] D. H. Wachsstock, W. H. Schwarz, and T. D. Pollard, Cross-linker dynamics determine the mechanical properties of actin gels, Biophys J. 66, 801 (1994).

[78] W. S. J. Xu, J. Kas, T. Stossel, P. Janmey, and T. Pollard, Mechanical properties of actin filament networks depend on preparation, polymerization conditions, and storage of actin monomers, Biophys. J. 74, 2731 (1998).

[79] H. Isambert and A. C. Maggs, Dynamics and rheology of actin solutions, Macromolecules 29, 1036 (1996).

[80] D. C. Morse, Tube diameter in tightly entangled solutions of semiflexible polymers, Phys. Rev. E 63, 031502 (2001).

[81] R. Everaers, S. K. Sukumaran, G. S. Grest, C. Svaneborg, A. Sivasubramanian, and K. Kremer, Rheology and microscopic topology of entangled polymeric liquids, Science 303, 823 (2004).

[82] C. Tzoumanekas and D. N. Theodorou, Topological analysis of linear polymer melts: A statistical approach, Macromolecules 39, 4592 (2006).

[83] F. Wagner, G. Lattanzi, and E. Frey, Conformations of confined biopolymers, Phys. Rev. E 75, 050902 (2007).

[84] S. Kaufmann, J. Ks, W. H. Goldmann, and G. Isenberg, Talin anchors and nucleates actin filaments - a direct demonstration at lipid membranes, FEBS Lett. 314, 203 (1992).

[85] L. L. Goff, O. Hallatschek, E. Frey, and F. Amblard, Tracer studies on f-actin fluctuations, Phys. Rev. Lett. 89, 258101 (2002).

[86] C. F. Schmidt, M. Baermann, G. Isenberg, and E. Sackmann, Chain dynamics, mesh size, and diffusive transport in networks of polymerized actin. a quasielastic light scattering and microfluorescence study, Macromolecules 22, 3638 (1989).

[87] H. Kleinert, J. Math. Phys. 27, 3003 (1986).

[88] F. G. Schmidt, B. Hinner, E. Sackmann, and J. X. Tang, Viscoelastic properties of semiflexible filamentous bacteriophage fd, Phys. Rev. E 62, 5509 (2000).

[89] M. Kawamura and K. Maruyama, Electron microscopic particle length of f-actin polymerized in vitro, Journal of Biochemistry 67, 437 (1970).

[90] M. A. Dichtl and E. Sackmann, Colloidal probe study of short time local and long time reptational motion of semiflexible macromolecules in entangled networks, New. J. Phys. 1, 18.1 (1999).

[91] M. Doi, Effect of chain flexibility on the dynamics of rodlike polymers in the entangled state, Journal of Polymer Science 73, 93 (1985).

[92] K. C. Holmes, D. Popp, W. Gebhard, and W. Kabsch, Atomic model of the actin filament, Nature 347, 44 (1990).

[93] H. Hinsch, J. Wilhelm, and E. Frey, Quantitative tube model for semiflexible polymer solutions, E. Phys. J. E 24, 35 (2007).

[94] N. Selve and A. Wegner, Rate of treadmilling of actin filaments in vitro, Journal of Molecular Biology 187, 627 (1986).

[95] W. K. Hastings, Monte carlo sampling methods using markov chains and their applications, Biometrika 57, 97 (1970).

[96] M. E. Fisher, Magnetism in one-dimensional systems - the heiseberg model for infinite spin, American Journal of Physics 32, 343 (1964).

[97] V. I. Manousiouthakis and M. W. Deem, Strict detailed balance is unnecessary in monte carlo simulation, J. Chem. Phys 110, 2753 (1999).

[98] A. Baumgartner and M. Muthukumar, A trapped polymer chain in random porous media, J. Chem. Phys. 87, 3082 (1987).

[99] M. E. Cates and R. C. Ball, Statistics of a polymer in a random potential, with implications for a nonlinear interfacial growth model, J. Phys. France 49, 2009 (1988).

[100] S. Edwards and M. Muthukumar, The size of a polymer in random media, J. Chem. Phys. 89, 2435 (1988).

[101] J. U. Sommer and A. Blumen, Nonmonotonic extension of polymers in aperiodic potentials, Phys. Rev. Lett. 79, 439 (1997).

[102] J. Xu, Y. Tseng, and D. Wirtz, Strain hardening of actin filament networks, J. Biol. Chem. 275, 35886 (2000).

[103] C. Semmrich, R. J. Larsen, and A. R. Bausch, Nonlinear mechanics of entangled f-actin solutions, Soft Matter 4, 1675 (2008).

[104] M. L. Gardel, J. H. Shin, F. C. MacKintosh, L. Mahadevan, P. Matsudaira, and D. A. Weitz, Elastic behavior of cross-linked and bundled actin networks, Science 304, 1301 (2004).

[105] M. Tassieri, R.M.L.Evans, L. Barbu-Tudoran, G.N.Khaname, J. Trinick, and T. A. Waigh, Dynamics of semiflexible polymer solutions in the highly entangled regime, Phys. Rev. Lett. 101, 198301 (2008).

[106] C. Heussinger and E. Frey, Floppy modes and nonaffine deformations in random fiber networks, Phys. Rev. Lett. 97, 105501 (2006).

[107] F. Höfling, T. Munk, E. Frey, and T. Franosch, Entangled dynamics of a stiff polymer, Phys. Rev. E 77, 060904 (R) (2008).

[108] H. Hinsch and E. Frey, Conformations of entangled semiflexible polymers, to be published (2009).

[109] W. H. Press, S. A. Teukolsky, W. T. Vetterling, and B. P. Flannery, editors, Numerical Recipes, Cambridge University Press, 2002.

[110] J. Nelder and R. Mead, A simplex method for function minimization, Comp. J. 7, 308 (1965).

[111] S. Aragon and R. Pecora, Dynamics of wormlike chains, Macromolecules 18, 1868 (1985).

Danksagung

An erster Stelle gebührt mein Dank Prof. Erwin Frey, der meine Promotion betreut hat. Ohne sein umfangreiches physikalisches Wissen und seine kreativen Ideen wäre diese Arbeit nicht möglich gewesen. Erwins Begeisterung für die biologische Physik haben mich immer wieder motiviert und er hat mir alle Freiheit und Unterstützung gegeben, die ich brauchte.

Eine große Hilfe war auch Dr. Claus Heussinger, auf dessen Wissen über Polymere ich nicht nur in unserer gemeinsamen Bürozeit, sondern auch bei meinen Besuchen in Lyon zurückgreifen konnte.

Für die fruchtbare Kollaboration möchte ich mich bei Prof. Rudolf Merkel und Marta Romanowska vom Forschungzentrum Jülich bedanken.

Finanziell wurde diese Arbeit vor allem durch das Internationale Doktorandenkolleg Nanobiotechnologie unterstützt. Besonders für die großzügigen Reisemittel möchte ich mich bedanken.

Ein verläßliche Hilfe für alle kleinen und großen Probleme habe ich stets bei den Mitgliedern des Lehrstuhls gefunden. Mein besonderer Dank geht an Claus und Felix, mit denen ich das Büro teilte, an Anna für ihre unterhaltsame Gesellschaft beim Rauchen und an Mark für all die gemeinsamen Feierabende.

Für das Korrekturlesen der Arbeit möchte ich mich bei Kathrin, Benedikt und Anna bedanken.

Ein Dankeschön auch an Kajetan, Sebastian, Sven und Philip, die mich immer wieder auf die andere Seite des Zaunes haben blicken lassen.

Besonderen Dank schulde ich meinen Mitbewohnerinnnen Sandra und Adriane für ihre herzliche Aufnahme und stete Gastfreundschaft.

Und vor allem danke ich PG276 für die unerschütterliche Liebe zur Physik.

Printed by Books on Demand GmbH, Norderstedt / Germany